Springer Theses

Recognizing Outstanding Ph.D. Research

For further volumes:
http://www.springer.com/series/8790

Aims and Scope

The series "Springer Theses" brings together a selection of the very best Ph.D. theses from around the world and across the physical sciences. Nominated and endorsed by two recognized specialists, each published volume has been selected for its scientific excellence and the high impact of its contents for the pertinent field of research. For greater accessibility to non-specialists, the published versions include an extended introduction, as well as a foreword by the student's supervisor explaining the special relevance of the work for the field. As a whole, the series will provide a valuable resource both for newcomers to the research fields described, and for other scientists seeking detailed background information on special questions. Finally, it provides an accredited documentation of the valuable contributions made by today's younger generation of scientists.

Theses are accepted into the series by invited nomination only and must fulfill all of the following criteria

- They must be written in good English.
- The topic should fall within the confines of Chemistry, Physics, Earth Sciences, Engineering and related interdisciplinary fields such as Materials, Nanoscience, Chemical Engineering, Complex Systems and Biophysics.
- The work reported in the thesis must represent a significant scientific advance.
- If the thesis includes previously published material, permission to reproduce this must be gained from the respective copyright holder.
- They must have been examined and passed during the 12 months prior to nomination.
- Each thesis should include a foreword by the supervisor outlining the significance of its content.
- The theses should have a clearly defined structure including an introduction accessible to scientists not expert in that particular field.

Takanori Sugiyama

Finite Sample Analysis in Quantum Estimation

Doctoral Thesis accepted by
The University of Tokyo, Tokyo, Japan

Author
Dr. Takanori Sugiyama
Department of Physics
Graduate School of Science
The University of Tokyo
Tokyo
Japan

Supervisor
Mio Murao
Department of Physics
Graduate School of Science
The University of Tokyo
Tokyo
Japan

ISSN 2190-5053 ISSN 2190-5061 (electronic)
ISBN 978-4-431-56183-5 ISBN 978-4-431-54777-8 (eBook)
DOI 10.1007/978-4-431-54777-8
Springer Tokyo Heidelberg New York Dordrecht London

Softcover reprint of the hardcover 1st edition 2014

Printed on acid-free paper

Springer is part of Springer Science+Business Media (www.springer.com)

Parts of this thesis have been published in the following journal articles:

1. Takanori Sugiyama, Peter S. Turner, Mio Murao, "Adaptive experimental design for one-qubit state estimation with finite data based on a statistical update criterion", Physical Review A 85, 052107 (2012).

2. Takanori Sugiyama, Peter S. Turner, Mio Murao, "Effect of non-negativity on estimation errors in one-qubit state tomography with finite data", New Journal of Physics 14, 085005 (2012).

3. Takanori Sugiyama, Peter S. Turner, Mio Murao, "Precision-Guaranteed Quantum Tomography", Physical Review Letters 111, 160406 (2013).

Supervisor's Foreword

Quantum mechanics has counterintuitive characteristics that distinguish it from classical mechanics. The probabilistic nature of measurement and the existence of nonlocal correlations are representative of such characteristics. Recent developments in technology that allow the fine manipulation of physical systems at quantum levels have opened up possibilities for breakthroughs in information technology, utilizing these characteristics. Checking the quantum properties of generated states and applied operations is crucial for the successful implementation of quantum manipulations in quantum information technology. In particular, the identification of generated states or the accuracy of applied operations by measuring quantum systems is called quantum tomography. For quantum tomography, due to the probabilistic nature of measurement in quantum mechanics, we need to perform measurements on many copies of the same system to obtain enough data to build useful statistics and we also need a scheme to estimate the states or operations by processing the data.

Quantum estimation refers to methods for estimating aspects of a quantum system from data obtained by measurements. The probabilistic nature of the measurement and the limited number of measurements mean that there are always errors in estimation results known as statistical errors. The size of the error depends on the choice of the estimation scheme. The key aims of quantum estimation theory are to evaluate the size of the errors as accurately as possible for a given estimation scheme, and to find schemes which give more precise estimates using fewer data points. In order to evaluate the size of statistical errors in real experiments, we need to analyze the behavior of statistical errors for a finite number of measurements. However, most theoretical of quantum estimation have been for the infinite case and there were few quantitative results for finite measurements that are applicable to experiments.

Dr. Takanori Sugiyama has developed quantum estimation theory with finite data in a way applicable to two quantum estimation problems: tests of Bell-type nonlocal correlations and quantum tomography. He used expected losses and estimation error probabilities as figures of merit for evaluating the size of the estimation errors. He presented evaluations of the estimation errors in tests of Bell-type nonlocal correlations and quantum tomography for finite data. He proposed a new estimator called the extended norm-minimization estimator. He derived upper bounds on estimation errors for extended linear, extended norm-minimization, and maximum-likelihood estimation schemes. Furthermore, he showed that the extended norm-minimization estimator provides precision-guaranteed quantum tomography. This was the first result directly applicable to evaluating the precisions of both estimation and preparation in quantum tomography. These results are important contributions to the field, giving us powerful tools for evaluating estimation errors in quantum information experiments.

He also presented an estimation scheme to improve estimation accuracy with adaptive design of experiments where successive measurements are updated according to previously obtained outcomes and measurement settings. He focused on one-qubit state estimations and derived an analytic solution to the update problem in the case of rank-1 projective measurements. Using this solution, he numerically compared the performances of adaptive and non-adaptive schemes for finite data sets, and showed that the adoptive scheme gives more precise estimates than standard quantum tomography.

Based on these achievements, Dr. Sugiyama was awarded the Student Research Prize in March 2013 by the Graduate School of Science, The University of Tokyo.

Tokyo, March 2013 Mio Murao

Acknowledgments

First, I would like to express my gratitude to my supervisor, Prof. Mio Murao, for all our discussions and her thoughtful and continuous support for the past 5 years. I would particularly like to thank my other supervisor, Prof. Peter S. Turner for all our discussions and his encouragement through the past 5 years.

I would like to thank my thesis examiners, Prof. Akira Shimizu, Prof. Makoto Gonokami, Prof. Masato Koashi, Prof. Yasunobu Nakamura, and Prof. Kimio Tsubono, for instructive comments on my thesis.

I am deeply grateful to Mrs. Yuki Amano, Mrs. Mami Hara, and Mrs. Yumiko Wada for their administrative support. I also express my appreciation to my colleagues Dr. Michal Hajdusek, Dr. Fabian Furrer, Dr. Akihito Soeda, Dr. Yoshifumi Nakata, Mr. Shojun Nakayama, Mr. Eyuri Wakakuwa, Mr. Seiseki Akibue, Mr. Kotaro Kato, Mr. Jisho Miyazaki, and Mr. Kosuke Nakago.

I would like to thank Dr. Fuyuhiko Tanaka for helpful discussion on mathematical statistics and Dr. Terumasa Tadano for useful advice on numerical simulation.

This work was supported by a JSPS Research Fellowship for Young Scientists (22-7564) and the Project for Developing Innovation Systems of the Ministry of Education, Culture, Sports, Science and Technology (MEXT), Japan. Two of my research trips in the Ph.D. course were supported by the Program for Fosterage of Internationally Active Young Physicist (Institutional Program for Young Researcher Overseas Visits, JSPS).

Finally, I would like to express my respect and appreciation to my family. After leaving home, I have realized that I received a very conscientious education from my parents and relatives.

Contents

Chapter 1
Introduction

Recent dramatic development of theory and experiment for microscopic systems has made it possible to investigate Nature more deeply and to utilize Her often surprising properties for our own purposes. Tests of quantum entanglement is a good early example. In 1935 Einstein, Podolsky, Rosen, and Schrödinger pointed out that quantum mechanics allows the existence of a counterintuitive correlation, now called entanglement, in quantum systems [1, 2]. In 1964 Bell proposed an experiment for testing its existence in practice [3], which motivated the proposal of many similar types of experiments [4]. Such methods are now called tests of Bell-type correlation. In 1981 the first experimental demonstration of such a test was done by Aspect, Grangier, and Roger in a photonic system, clearly establishing the existence of entanglement [5]. Today many experiments for such tests have been done in several types of physical systems, including photons [6], neutrons [7], trapped ions [8, 9], superconducting qubits [10], NV-centers in a diamond [11], and B mesons [12]. This has driven the recent growth of the field of quantum information, with quantum cryptography and quantum computation as good later examples. In 1984 and 1991, Bennett and Brassard, and Ekert independently proposed a cryptographic protocol using quantum systems [13, 14]. Surprisingly, the protocol is secure in principle for the ideal case, and the security of nonideal cases was proved later [15, 16]. In 1994 Shor proposed a computational algorithm for factorizing in prime integers using quantum systems [17]. It is again surprising that the protocol runs in polynomial time, something widely believed to be impossible for classical algorithms. With this cryptographic protocol and computational algorithm which outperform their classical counterparts as motivation, several new protocols and algorithms have been proposed [18, 19], with experimentalists in diverse fields working to implement many of them in their labs [20, 21].

Protocols in quantum cryptography and algorithms in quantum computation are examples of quantum information protocols. When experimentalists try to implement a quantum information protocol in their lab, they need to confirm the successful preparation of specific quantum states and operations. The standard method used for this verification is called quantum tomography [22]. For simplicity, let us consider

T. Sugiyama, *Finite Sample Analysis in Quantum Estimation*, Springer Theses,
DOI: 10.1007/978-4-431-54777-8_1,

quantum state tomography, where, we have many identical copies of an unknown quantum state, and our purpose is to identify the state by performing predetermined measurements on the copies. We then try to estimate the true state from the data obtained. In quantum mechanics, measurement outcomes are obtained probabilistically. Data obtained in quantum state tomography includes two types of errors. One is caused by statistical fluctuation, and the other is due to systematic noise. The former is called statistical error, and the latter is called systematic error. Because of these errors, an estimated state does not coincide with the true state. The difference between the estimated and true state is called an estimation error. In quantum state tomography, it is very important to accurately evaluate the size of the estimation error in order to verify a successful preparation of a quantum state. This importance of accurately evaluating estimation error is also true in the test of Bell-type correlations. In an experimental Bell-type test, we perform measurements on a bipartite system and calculate a correlation function from the outcomes obtained. In this case, the estimation object is the value of the correlation function. If the value is larger than a certain threshold, it indicates that there exists a counterintuitive correlation between the two systems. The estimated value, of course, includes estimation errors. In order to verify the existence of the correlation, we need to evaluate the size of the estimation error and show that it is sufficiently small. In general, to estimate from data fluctuating probabilistically and affected by noise, and to evaluate the size of estimation error are topics in statistical estimation theory. Statistical estimation for quantum systems is called quantum estimation. Usually, the effect of the systematic error is approximated by introducing a model, and is assumed to be known. Therefore, the analysis of the estimation error is usually reduced to that of the statistical error.

In statistical estimation, how to choose the measurements used for the estimation is called an experimental design, and how to calculate the estimation results from the obtained data is called an estimator [23]. It is a key aim of both classical and quantum estimation theory to find a combination of experimental design and estimator which gives us more precise estimation results using fewer measurement trials. Before trying to find better combinations, we need to decide how to evaluate the size of the estimation error for a given combination. In statistical estimation theory, there are two main ways to evaluate the size of estimation errors. One is to use expected loss. We evaluate the difference between the true object and the estimate by what is called a loss function, which we average by taking its statistical expectation. The result is called the expected loss. Mean squared error and variance are the examples of expected loss, and an expected loss is a generalization of error bars. The other way is to use error probability. Like with expected loss, we evaluate the difference between the true object and the estimate by a loss function, but now calculate the probability that we obtain data that gives a difference larger than some threshold. An error probability tells us the probability of obtaining estimation results that deviate from the truth by more than some amount. An expected loss tends to be used for evaluating the rough size of an estimation error, while an error probability is used to more accurately evaluate its size. In quantum estimation theory, there are results for expected losses and error probabilities for infinitely many data. These constitute what is called the large sample theory or the asymptotic theory of quantum estimation [24]. In real experiments, however, the amount of available data is finite. Compared to the asymptotic analysis, it is more difficult to analyze the size of estimation errors for finite data. Theoretical methods for accurately evaluating estimation errors in tests of Bell-type correlations and in quantum tomography with finite data have not been established. In this thesis we develop the theory of statistical parameter estimation toward a finite sample theory applicable to such real world estimation problems.

In classical statistics, improving the estimation precision is as important a topic as accurately evaluating estimation errors. This is also true in quantum estimation. Suppose that the mathematical model of our estimation object and the way we evaluate estimation error are determined. When we try to improve the estimation precision, we can choose to change the experimental design or the estimator. Let us consider the case in which we try to improve the experimental design. In general the optimal design of experiment depends on the true value of the estimation object. However, we do not know the true value and that is our motivation for estimation. This lack of knowledge of the true value forces us to choose a sub-optimal design of experiment. Adaptive design of experiments is one such sub-optimal design. Suppose that a set of identical copies of an unknown estimation object is given, and we perform a measurement on each copy. The recipe for adaptive design of experiments is as follows. We perform a measurement on the first copy and obtain an outcome. From that outcome, we calculate an estimate. Before performing the next measurement, we tune the measurement apparatus in the optimal way for the estimated value. That is, we treat the estimate as the true value temporarily and tune the apparatus accordingly. Afterwards, we perform the tuned measurement on the second copy, calculate an estimate from the first and second outcomes, and repeat the tuning, measurement, and estimation until we reach the final copy. The class of adaptive designs of experiments clearly includes nonadaptive and independent designs of experiments, and so is expected to outperform them. In this thesis, we propose an adaptive design of experiment for one-qubit state estimation and show that our method gives more accurate estimation results than standard quantum state tomography.

We summarize the contents above. In this thesis we analyze statistical estimation errors in tests of Bell-type correlations and in quantum tomography for finite samples. Chapters 2 and 3 are devoted to preliminary. In Chap. 2, we review the postulates of quantum mechanics used in quantum information and quantum estimation. We also explain the importance of finite sample analysis of estimation errors in the test of Bell-type correlations and quantum tomography. In Chap. 3, we introduce the basic concepts and notation in statistical parameter estimation theory and show the known results necessary for analyzing errors in quantum estimation. Chapters 4–6 treat our results on the accurate evaluation and an improvement of estimation errors. In Chap. 4, we analyze expected losses and error probabilities in the test of Bell-type correlations with finite samples. We derive explicit forms of their upper bounds that are directly applicable for evaluating the validity of the violation of the CHSH inequality in experiments. In Chap. 5, we analyze expected losses and error probabilities in quantum tomography with finite samples. We focus on three estimators, extended linear, extended norm-minimization, and maximum-likelihood, and derive

Table 1.1 Ways to evaluate estimation errors and the sections of this thesis discussing the corresponding results

	Test of bell-type correlations	Quantum tomography	Adaptive design of experiments
Expected loss	Sect. 4.3.2	Sect. 5.2	Sect. 6.2
Error probability	Sect. 4.3.3	Sect. 5.3	–

upper bounds on their estimation errors. In Chap. 6, we consider an adaptive design of experiment in one-qubit state estimation. We focus on a measurement update criterion called the A-optimality criterion, and numerically evaluate its performance. Our numerical results indicate that the A-optimal design of experiment gives more precise estimation results than standard quantum state tomography. Table 1.1 gives the sections containing our results for each of these topics. In Chap. 7, we summarize the results of this thesis.

References

1. A. Einstein, B. Podolsky, N. Rosen, Phys. Rev. **47**, 777 (1935). doi:10.1103/PhysRev.47.777
2. E. Schrödinger, in *Mathematical Proceedings of the Cambridge Philosophical Society*, vol. 31 (1935), p. 555. doi:10.1017/S0305004100013554
3. J.S. Bell, Physics **1**, 195 (1964)
4. N. Gisin, arXiv:070221[quant-ph] (2007)
5. A. Aspect, P. Grangier, G. Roger, Phys. Rev. Lett. **47**, 460 (1981). doi:10.1103/PhysRevLett.47.460
6. A. Zeilinger, Rev. Mod. Phys. **71**, S288 (1999). doi:10.1103/RevModPhys.71.S288
7. Y. Hasegawa, R. Loldl, G. Badurek, M. Baron, H. Rauch, Nature **425**, 45 (2003). doi:10.1038/nature01881
8. M.A. Rowe, D. Klelplnskl, V. Meyer, C.A. Sackett, W.M. Itano, C. Monroe, D.J. Wineland, Nature **409**, 791 (2001). doi:10.1038/35057215
9. D.N. Matsukevich, P. Maunz, D.L. Moehring, S. Olmschenk, C. Monroe, Phys. Rev. Lett. **100**, 150404 (2008). doi:10.1103/PhysRevLett.100.150404
10. M. Ansmann, H. Wang, R.C. Bialczak, M. Hofheinz, E. Lucero, M. Neeley, A.D. O'Connell, D. Sank, M. Weides, J. Wenner, A.N. Cleland, J.M. Martinis, Nature **461**, 504 (2009). doi:10.1038/nature08363
11. G. Waldherr, P. Neumann, S.F. Huelga, F. Jelezko, J. Wrachtrup, Phys. Rev. Lett. **107**, 090401 (2011). doi:10.1103/PhysRevLett.107.090401
12. A. Go, J. Mod. Phys. **51**, 991 (2004). doi:10.1080/09500340408233614
13. C.H. Bennett, G. Brassard, in *Proceedings of IEEE International Conference on Computers, Systems and Signal Processing* (IEEE Press, 1984), p. 175
14. A.K. Ekert, Phys. Rev. Lett. **67**, 661 (1991). doi:10.1103/PhysRevLett.67.661
15. P.W. Shor, J. Preskill, Phys. Rev. Lett. **85**, 441 (2000). doi:10.1103/PhysRevLett.85.441
16. D. Mayers, J. ACM **48**, 351 (2001). doi:10.1145/382780.382781
17. P.W. Shor, in *Proceedings of the 35th Annual Symposium on Foundations of Computer Science* (IEEE Press, 1994). doi:10.1109/SFCS.1994.365700
18. S. Jordan, Quantum algorithm zoo. http://math.nist.gov/quantum/zoo/

19. V. Scarani, H. Bechmann-Pasquinucci, N.J. Cerf, M. Dušek, N. Lütkenhaus, M. Peev, Rev. Mod. Phys. **81**, 1301 (2009). doi:10.1103/RevModPhys.81.1301
20. W.P. Schleich, H. Walther (eds.), *Elements of Quantum Information* (WILEY-VCH, Weinheim, 2007)
21. T.D. Ladd, F. Jelezko, R. Laflamme, Y. Nakamura, C. Monroe, J.L. O'Brien, Nature **464**, 45 (2010). doi:10.1038/nature08812
22. M. Paris, J. Řeháček (eds.), *Quantum State Estimation*, Lecture Notes in Physics (Springer, Berlin, 2004)
23. E.L. Lehmann, *Theory of Point Estimation*, Springer Texts in Statistics (Springer, New York, 1998)
24. M. Hayashi (ed.), *Asymptotic Theory of Quantum Statistical Inference: Selected Papers* (World Scientific, Singapore, 2005)

Chapter 2
Quantum Mechanics and Quantum Estimation: Background and Problems in Quantum Estimation

2.1 Quantum Mechanics: Operational Approach

Quantum mechanics is a theory for describing measurement results on microscopic systems. The description consists of two parts, the measured system and the measurement apparatus. In quantum mechanics, the measured systems are microscopic systems like atoms, molecules, weak electromagnetic waves, and so on. Experimental results depend on the state of the system and the action of the measurement. When we perform measurement on a microscopic system, the measurement outcomes obtained are probabilistic. Quantum mechanics gives us the rule for calculating the probability distribution. As time passes or when we perform measurement, the state of the system changes. Quantum mechanics also gives us the rule for this change.

In this thesis, we use the postulates of quantum mechanics as they are widely used in quantum information theory [1, 2]. These consist of four statements. We consider finite dimensional systems only, and the postulates introduced below are a limited version valid for finite dimensional systems. Brief explanations follow the statements below.

Postulate (Quantum Mechanics)

1. System and state
   For an arbitrary system, there exists at least one corresponding Hilbert space $\mathscr{H}$. The state of the system is characterized by a positive semidefinite and unit trace matrix on the Hilbert space, $\hat{\rho}$.
2. Measurement (probability distribution)
   For an arbitrary measurement, there exists a corresponding set of matrices $\boldsymbol{\Pi} = \{\hat{\Pi}_x\}_{x \in \mathscr{X}}$ where $\hat{\Pi}_x$ are positive-semidefinite matrices on $\mathscr{H}$ satisfying

$$\sum_{x \in \mathscr{X}} \hat{\Pi}_x = \hat{\mathbb{1}}. \tag{2.1}$$

T. Sugiyama, *Finite Sample Analysis in Quantum Estimation*, Springer Theses,

DOI: 10.1007/978-4-431-54777-8_2,

When we perform a measurement characterized by $\boldsymbol{\Pi}$, the probability we obtain an outcome x is

$$p(x|\boldsymbol{\Pi}, \hat{\rho}) = \mathrm{Tr}[\hat{\Pi}_x \hat{\rho}]. \tag{2.2}$$

3. Composite system
   If there are systems characterized by Hilbert spaces $\mathscr{H}$ and $\mathscr{H}'$ respectively, the total system is characterized by the tensor product of these Hilbert spaces, $\mathscr{H} \otimes \mathscr{H}'$.
4. State transition
   a. Deterministic transition
      Suppose that the state transition is not induced by a measurement and the state is $\hat{\rho}$ at time t. For any $t' > t$, there exists a linear, trace-preserving, and completely positive (TPCP) map κ satisfying

$$\hat{\rho}' = \kappa(\hat{\rho}). \tag{2.3}$$

   b. Probabilistic transition
      A quantum measurement is a set of linear and completely positive (CP) maps $\boldsymbol{\kappa} = \{\kappa_x\}_{x \in \mathscr{X}}$ such that $\sum_{x \in \mathscr{X}} \kappa_x$ is trace-preserving (TP). Suppose that the state just before the measurement is characterized by $\hat{\rho}$. When we obtain an outcome x, the state just after the measurement is given by

$$\hat{\rho}_x = \frac{\kappa_x(\hat{\rho})}{\mathrm{Tr}[\kappa_x(\hat{\rho})]} \tag{2.4}$$

      and the probability we obtain an outcome x is given as

$$p(x|\boldsymbol{\kappa}, \hat{\rho}) = \mathrm{Tr}[\kappa_x(\hat{\rho})]. \tag{2.5}$$

A system obeying the postulates is called a quantum system. A positive semidefinite and unit trace matrix is called a density matrix. A quantum state is characterized by a density matrix. Let $\mathscr{S}(\mathscr{H})$ denote the set of all density matrices on $\mathscr{H}$. When a density matrix is written as $\hat{\rho} = |\psi\rangle\langle\psi|$, the state is called pure, and $|\psi\rangle$ is called a state vector. In general a state described by a density matrix can be interpreted as a probabilistic mixture of pure states. Thus a density matrix is a generalization of a state vector description.

A set of positive semidefinite matrices satisfying Eq. (2.1) is called a positive operator-valued measure (POVM). Let $\hat{A}$ denote an Hermitian matrix on $\mathscr{H}$ and $\hat{A} = \sum_{x \in \mathscr{X}} a_x \hat{E}_x$ be the eigenvalue decomposition. Then the set $\{\hat{E}_x\}_{x \in \mathscr{X}}$ is a POVM, and the corresponding measurement is called the projective measurement of $\hat{A}$.

When a system is isolated from any other systems, the deterministic state transition is characterized by an unitary matrix. Generally, a system of interest in quantum information is not isolated. When a state transition occurs deterministically, the state transition is called a quantum process. A map κ on $\mathscr{S}(\mathscr{H})$ is called trace-preserving if it satisfies $\mathrm{Tr}[\kappa(\hat{\rho})] = 1$ for any $\hat{\rho} \in \mathscr{S}(\mathscr{H})$. A map κ is called positive if it satisfies that $\kappa(\hat{\rho})$ is positive semidefinite for any $\hat{\rho} \in \mathscr{S}(\mathscr{H})$. Let $\mathscr{H}'$ denote a d'-dimensional Hilbert space and $\iota_{\mathscr{H}'}$ denote the identity map on $\mathscr{S}(\mathscr{H}')$. A map κ on $\mathscr{S}(\mathscr{H})$ is called completely positive if the composite map $\kappa \otimes \iota_{\mathscr{H}'}$ is positive for any $\hat{\rho} \in \mathscr{S}(\mathscr{H} \otimes \mathscr{H}')$ and any positive integer d'. It is known that a quantum process is characterized by a linear TPCP map.

When a state transition occurs probabilistically, the state after the transition depends on a probabilistic outcome. It is known that such transitions are characterized by a set of linear and completely positive maps whose sum is trace-preserving. Such a set is called a quantum instrument.

POVMs and quantum instruments both describe the actions of quantum measurements. For a given quantum measurement, the corresponding POVM characterizes the probability distribution and the quantum instrument does the state transition. In this thesis, we treat quantum estimation and analyze estimation errors. In quantum estimation, we perform measurements on a quantum system in order to know some aspect of the system. In the analysis of the estimation errors, the probability distribution of the measurement plays an important role, and we will frequently use Eq. (2.2) in the following chapters.

2.2 Quantum Estimation

Suppose that there is a quantum system of interest and we perform quantum measurements on the system. Quantum estimation is a term for estimating some aspect of the system from our knowledge of the measurement apparatus and the outcomes obtained. By definition, quantum estimation is related to almost all experiments on quantum systems. Due to the probabilistic behavior of the measurement outcomes and the finiteness of the number of measurement trials, there always exist errors in our estimation results. The size of the error depends on the choice of measurements and the estimation procedure. In statistics, the former is called an experimental design while the latter is called an estimator. Key aims of both classical and quantum estimation theory are to evaluate the size of the error as accurately as possible for a given combination of experimental design and estimator, as well as to find a combination of experimental design and estimator which gives us more accurate estimation results using fewer measurement trials.

In this thesis we focus on two quantum estimation problems, namely tests of Bell-type correlations and quantum tomography. Here we give brief reviews of these estimation problems, and we explain the details in the corresponding chapters later.

1. Test of Bell-type correlations
   Bell-type correlation is a class of correlations between two systems. Here the systems are not necessary quantum. The correlation depends upon a set of measurement outcomes and probability distributions of measurements on the

two systems. One of the most famous examples is the CHSH correlation. The value of the CHSH correlation tells us whether the probability distributions can be described by a local hidden-variable model or not, where, roughly speaking, a local hidden variable model corresponds to classical mechanics. When the probability distributions can be described by a local hidden-variable model, the value of the CHSH correlation must be smaller than or equal to 2. On the other hand, it is known that when the probability distributions obey quantum mechanics, the value of the CHSH correlation can be up to $2\sqrt{2}$. If we observe a value of CHSH correlation larger than 2, it proves that there exists systems which cannot be described by any local hidden-variable model. Therefore tests of the CHSH correlation play a very important role in the foundation of physics, and the estimation error of the value should be relatively small. In Chap. 4, we evaluate estimation errors of Bell-type correlations accurately.

2. Quantum tomography
   Quantum tomography is a general term for methods used to completely determine the mathematical representation of tomographic objects, explained below. Tomographic objects fall into four categories: quantum states, quantum processes, probability distributions of quantum measurement outcomes, and state transitions caused by quantum measurements. From the postulates introduced in the previous section, the corresponding mathematical representations are density matrices, linear CPTP maps, POVMs, and quantum instruments, respectively.

   In quantum information, quantum tomography is used for verifying a successful preparation of quantum states or processes that will be used for a quantum information protocol. Evaluating the estimation error accurately is very important because the size of the estimation error affects the performance of the protocol. In Chap. 5, we evaluate estimation errors in quantum tomography accurately, and in Chap. 6 we introduce an experimental design which can improve estimation error.

2.3 Problems in Quantum Estimation

As explained in the previous section, it is important to evaluate the size of estimation errors accurately in quantum estimation. Nevertheless, such evaluation has not been done in quantum information experiments so far. The author believes that there are two reasons for this lack of the evaluation. First, in the past experiments, the size of the systematic noise was large and considered dominant compared to the effect of statistical errors. However, as experimental techniques becomes better, the size of systematic noise becomes smaller, and nowadays there are experiments such that statistical errors cannot be neglected. Second, in order to evaluate the size of statistical errors in real experiments, we need to analyze the behavior of statistical errors for a finite number of measurement trials. Theoretically, it is more difficult to analyze statistical errors for finite trials than for infinite one. There are a lot of theoretical studies on the analysis for the infinite case, and these are called the asymptotic theory of quantum estimation. However there are no quantitative results on finite measurements which are applicable for quantum information experiments. Experiments could not evaluate the size of statistical errors even if they wanted to do so because the theoretical tools for evaluating statistical errors are not available.

In this thesis, we develop quantum estimation theory to make it applicable for quantum information experiments with a finite number of measurement trials. Before this development, we need to understand how to treat estimation errors in quantum estimation mathematically. In Chap. 3, we introduce fundamental concepts and previously known results from statistical estimation theory.

References

1. M.A. Nielsen, I.L. Chuang, *Quantum Computation and Quantum Information* (Cambridge University Press, Cambridge, 2000)
2. M.M. Wilde, *Quantum Information Theory* (Cambridge University Press, Cambridge, 2013)

Chapter 3
Mathematical Statistics: Basic Concepts and Theoretical Tools for Finite Sample Analysis

3.1 Preliminaries

In this section, we explain the terminology of probability theory and statistical parameter estimation theory. We use a notation $A := B$ or $B =: A$ as the meaning that we define A by B. Let $\mathscr{A}$ denote a set and f be a function from $\mathscr{A}$ to $\mathbb{R}$. We use $\max_{a \in \mathscr{A}} f(a)$ and $\min_{a \in \mathscr{A}} f(a)$ as the maximal and minimal values of f over $\mathscr{A}$, respectively. We also use $\mathrm{argmax}_{a \in \mathscr{A}} f(a)$ and $\mathrm{argmin}_{a \in \mathscr{A}} f(a)$ as the elements maximizing and minimizing f over $\mathscr{A}$.

3.1.1 Probability Theory

In this subsection, we explain basic concepts in probability theory. In usual mathematical texts, probability theory is axiomatically introduced by using measure theory. However, we do not adopt the axiomatic approach, because in this thesis we treat only two simple probability distributions, multinomial and Gaussian distributions.

Let Ω denote a set of indices with M elements, i.e., $\Omega = \{1, \ldots, M\}$. A random variable X is defined as a function from Ω to $\mathbb{R}^k$, where k is an arbitrary positive integer. The set of the values of the random variable $\mathscr{X} := \{X(1), \ldots, X(M)\}$ is called the sample space of X. A probability distribution $\boldsymbol{p} = \{p(x)\}_{x \in \mathscr{X}}$ is defined as a set of M non-negative numbers satisfying $\sum_{x \in \mathscr{X}} p(x) = 1$. Let $\mathscr{P}(\mathscr{X})$ denote the set of all probability distributions on the sample space $\mathscr{X}$. When we observe a random variable X and obtain $x \in \mathscr{X}$ with probability $p(x)$, $\boldsymbol{p}$ is called the probability distribution of X. We use the notation $p(x|X)$ or $p_X(x)$ when we want to emphasize which random variable corresponds to a probability distribution (Table 3.1).

A subset E of $\mathscr{X}$ is called an event of X. The probability measure of X is a function from the events to $[0, 1] \in \mathbb{R}$ and defined as

T. Sugiyama, *Finite Sample Analysis in Quantum Estimation*, Springer Theses,
DOI: 10.1007/978-4-431-54777-8_3,

Table 3.1 Notation regarding random variables

Notation	Explanation
X	A random variable
$\mathscr{X}$	The sample space of X
x	An outcome of the random variable X
$p_X(x)$, $p(x\|X)$	The probability we observe X takes outcome x
$\boldsymbol{p}_X$, $\boldsymbol{p}(\cdot\|X)$	The probability distribution of X
$\mathbb{P}_X$	The probability measure of X
$\mathbb{E}[X]$	The expectation of X
$\mathbb{V}[X]$	The variance of X
$\sigma[X]$	The standard deviation of X

$$\mathbb{P}[E] := \sum_{x \in E} p(x). \tag{3.1}$$

Obviously a probability measure satisfies $\mathbb{P}[E] \geq 0$ and $\mathbb{P}[\mathscr{X}] = 1$. For an event E, $\mathbb{P}[E]$ is the probability that we obtain an outcome included in E when we observe the random variable X. As with probability distributions, we use the notation $\mathbb{P}[E|X]$ or $\mathbb{P}_X[E]$ when we emphasize the corresponding random variable. In this thesis, we mainly treat discrete and finite sample spaces, but in Sect. 5.2.3 we treat a continuous Gaussian distribution. In this case, Ω is not a discrete set but $\mathbb{R}^k$, and the sum in Eq. (3.1) is replaced by an integral.

As explained in Sect. 2.1, when we perform a measurement $\boldsymbol{\Pi} = \{\hat{\Pi}_x\}_{x \in \mathscr{X}}$ on a quantum system in a state $\hat{\rho}$, the probability distribution is given by

$$p(x|\boldsymbol{\Pi}, \hat{\rho}) = \mathrm{Tr}[\hat{\Pi}_x \hat{\rho}]. \tag{3.2}$$

The POVM $\boldsymbol{\Pi}$ corresponds to a random variable X, and the probability distribution is characterized by the pair $(\boldsymbol{\Pi}, \hat{\rho})$.

We call

$$\mathbb{E}[X] := \sum_{x \in \mathscr{X}} p_X(x)\, x \tag{3.3}$$

the expectation of a random variable X,

$$\mathbb{V}[X] := \mathbb{E}[X^2] - \mathbb{E}[X]^2 \tag{3.4}$$

the variance of X, and $\sigma[X] := \sqrt{\mathbb{V}[X]}$ the standard deviation of X.

Table 3.2 Notations regarding estimation object

Notation	Explanation
Θ	A parameter space
θ	A parameter in Θ
$p_{X,\theta}(x)$, $p(x\|X,\theta)$	A parametrized probability
$\boldsymbol{p}_{X,\theta} = \{p_{X,\theta}(x)\}$	A parametrized probability distributions
$\mathscr{M}$	A set of random variables
$\mathscr{P}_\theta = \{\boldsymbol{p}_{X,\theta}\}_{X\in\mathscr{M}}$	A set of parametrized probability distributions
$\mathscr{P}_\Theta = \{\mathscr{P}_\theta\}_{\theta\in\Theta}$	A statistical model

3.1.2 Statistical Parameter Estimation

Statistical parameter estimation is a general term for methods used to estimate a parameter characterizing a family of probability distributions from a set of observation results. A setting in statistical parameter estimation is characterized by a quintuplet of a statistical model, a number of observations, an experimental design, an estimator, and a figure of merit. In this subsection, we explain these basic concepts in the context of statistical parameter estimation, which are helpful in analyzing several mathematical problems in quantum estimation.

1. Estimation object (Table 3.2)
   Let θ denote a parameter in a parameter space $\Theta \subset \mathbb{R}^k$. This parameter space Θ is the set of candidate estimates in our estimation problem. Let $\mathscr{M}$ be a family of random variables X. Let us consider a set of probability distributions $\mathscr{P}_\theta := \{\boldsymbol{p}_{X,\theta}\}_{X\in\mathscr{M}}$, where $\boldsymbol{p}_{X,\theta} := \{p(x|X,\theta)\}_{x\in\mathscr{X}}$ is a parametrized probability distribution. The set $\mathscr{P}_\Theta := \{\mathscr{P}_\theta\}_{\theta\in\Theta}$ is called a statistical model.
   Let $\mathscr{P} := \{\boldsymbol{p}_X\}_{X\in\mathscr{M}}$ denote the set of probability distributions describing possible observations. In statistical parameter estimation, we assume that $\mathscr{P}$ is included in the statistical model $\mathscr{P}_\Theta$. The parameter θ satisfying $\mathscr{P}_\theta = \mathscr{P}$ is called the true parameter, and let θ_0 denote this true parameter. The purpose of statistical parameter estimation is to identify θ_0 from observation results.
   A parameter space Θ is called identifiable if it satisfies $\mathscr{P}_\theta \neq \mathscr{P}_{\theta'}$ for any $\theta \neq \theta'$ in Θ. If Θ is not identifiable, it is called unidentifiable. If Θ is unidentifiable, there exist parameters which cannot be identified by any observation chosen from $\mathscr{M}$ because their statistical properties are completely equivalent. We assume that the parameter space Θ is identifiable in what follows.
2. Experimental design (Table 3.3)
   Suppose that a statistical model $\mathscr{P}_\Theta$ is given. When we perform observations N times, we need to decide which random variable $X_n \in \mathscr{M}$ is observed in the n-th trial ($n = 1, 2, \ldots, N$). An experimental design (or design of experiment) is a sequence of random variables, $X^{\mathrm{exp}} := \{X_1, X_2, \ldots, X_n, \ldots\}$, and this is a recipe to decide which random variable will be observed at each trial. Let $\mathscr{X}_n$ denote the sample space of X_n and X^N denote the random variable sequence from

Table 3.3 Notation regarding experimental design

Notation	Explanation
N	The total number of observation trials
$n = 1, \ldots, N$	The index of the trials
X_n	A random variable observed at n-th trial
$\mathscr{X}_n$	The sample space of X_n
x_n	An outcome of the random variable X_n
$X^n = \{X_1, \ldots, X_n\}$	A random variable sequence up to the n-th trial
$\mathscr{X}^n = \mathscr{X}_1 \times \cdots \times \mathscr{X}_n$	The sample space of X^n
$\boldsymbol{x}^n = \{x_1, \ldots, x_n\}$	An outcome sequence of X^n
$X^{\mathrm{exp}} = \{X_1, \ldots X_n, \ldots\}$	An experimental design

$n = 1$ to $n = N$, i.e., $X^N := \{X_1, \ldots, X_N\}$. When X^N is independent and all random variables are same, i.e., $X_1 = \cdots = X_N$, the sequence X^N is called an independent and identically distributed (i.i.d.) random variable sequence. When X^N is i.i.d. for any N, we call X^{exp} an i.i.d. experimental design.

If an experimental design X^{exp} satisfies the following condition, we say that the parameter space Θ is identifiable by the experimental design X^{exp} with N samples.

Condition 3.1 (Identifiability)
For any $\theta,\ \theta' \in \Theta\ \theta \neq \theta'$,

$$\boldsymbol{p}_{X^N,\theta} \neq \boldsymbol{p}_{X^N,\theta'} \tag{3.5}$$

holds.

If X^{exp} does not satisfy Condition 3.1, i.e., there exists a pair of parameters $\theta \neq \theta'$ such that $\boldsymbol{p}_{X^N,\theta} = \boldsymbol{p}_{X^N,\theta'}$, Θ is called unidentifiable by X^{exp} with N samples. If Θ is unidentifiable by X^{exp} with N samples, there exists parameters which cannot be uniquely identified by any observation results of X^N. We assume that an experimental design X^{exp} satisfies Condition 3.1 in what follows.

3. Estimator (Table 3.4)
Suppose that we perform an experimental design X^{exp} and obtain a set of results $\boldsymbol{x}^N = \{x_1, \ldots, x_N\}$, $x_n \in \mathscr{X}_n$. We call a pair of an observed random variable and an obtained outcome, $D_n := (X_n, x_n)$, the N-th datum and a sequence $D^n := \{D_1, \ldots, D_n\}$ the data sequence up to the n-th trial. Our goal is to output a point in the parameter space from the knowledge of the observed random variables X^n and obtained outcomes $\boldsymbol{x}^n$. An estimator $\theta^{\mathrm{est}} := \{\theta_1^{\mathrm{est}}, \ldots, \theta_n^{\mathrm{est}}, \ldots\}$ gives such a calculation recipe. Each element of an estimator θ_n^{est} is a map from $\times_{i=1}^n(\mathscr{M}, \mathscr{X}_i)$ to $\mathbb{R}^k$. An estimate $\theta_n^{\mathrm{est}}(D^n)$ is called the n-th estimate. When the observed random

Table 3.4 Notation regarding estimators

Notation	Explanation
$D_n = (X_n, x_n)$	An n-th datum
$D^n = \{D_1, \ldots, D_n\}$	A data sequence up to the n-th trial
θ_n^{est}	A map from a data sequence to a parameter
$\theta_n^{\mathrm{est}}(D^n)$	An n-th estimate
$\theta^n = \{\theta_1^{\mathrm{est}}, \ldots, \theta_n^{\mathrm{est}}, \ldots\}$	An estimator

variables are clear, we omit the random variable dependency and use the notation $\theta_n^{\mathrm{est}}(\boldsymbol{x}^n)$ for the estimate. Sometimes we also omit the outcome dependency and write θ_n^{est} as the n-th estimate (rather than the n-th component of an estimator).

Example 3.1 (Linear estimator)
Suppose that $\mathscr{X}$ is finite and $X^{\exp}$ is i.i.d.. Let us consider the case in which the true parameter is represented as

$$\theta_0 = \sum_{x \in \mathscr{X}} a(x) p(x|X, \theta_0) \tag{3.6}$$
$$= \mathbb{E}_{\theta_0}[a(X)], \tag{3.7}$$

i.e., our estimation object is the expectation of a random variable a(X). Let $N_x(\boldsymbol{x}^N)$ denote the number of times that outcome x occurs in $\boldsymbol{x}^N$. A linear estimator θ^{L} is defined as the arithmetic mean of $a(X)$:

$$\theta_N^{\mathrm{L}}(\boldsymbol{x}^N) := \frac{1}{N} \sum_{n=1}^{N} a(x_n) \tag{3.8}$$
$$= \sum_{x \in \mathscr{X}} a(x) \nu_{N,x}(\boldsymbol{x}^N) \tag{3.9}$$
$$= \boldsymbol{a} \cdot \boldsymbol{\nu}_N(\boldsymbol{x}^N), \tag{3.10}$$

where

$$\nu_{N,x}(\boldsymbol{x}^N) := \frac{N_x(\boldsymbol{x}^N)}{N} \tag{3.11}$$

is called the relative frequency of x in $\boldsymbol{x}^N$. We analyze a linear estimator in a test of the CHSH inequality in Sects. 4.3.2 and 4.3.3, as well as in quantum tomography in Sects. 5.2.1 and 5.3.1.

Example 3.2 (Maximum likelihood estimator)
A maximum-likelihood estimator θ^{ML} is defined as

$$\theta_N^{\mathrm{ML}}(\boldsymbol{x}^N) := \underset{\theta\in\Theta}{\operatorname{argmax}}\ p(\boldsymbol{x}^N | X^N, \theta) \tag{3.12}$$

$$= \underset{\theta\in\Theta}{\operatorname{argmax}} \prod_{n=1}^{N} p(x_n | X_n, \theta). \tag{3.13}$$

We analyze a maximum-likelihood estimator in quantum tomography in Sects. 5.2.3 and 5.3.3, as well as in adaptive design of experiments in Sect. 6.2.2.

4. Loss function
A quadruplet $(\mathscr{P}_\Theta, N, X^{\mathrm{exp}}, \theta^{\mathrm{est}})$ specifies an estimation scheme. Suppose that an estimation scheme $(\mathscr{P}_\Theta, N, X^{\mathrm{exp}}, \theta^{\mathrm{est}})$ is given. In general, estimates $\theta_N^{\mathrm{est}}(D^N)$ can be different from the true parameter θ_0 for any finite N, because of the statistical fluctuation and systematic noise on observation results. For a given experimental design and estimator, it is important to evaluate the amount of that difference in order to guarantee the validity of the estimation results. The difference between $\theta_N^{\mathrm{est}}(D^N)$ and θ_0 is called the estimation error. Estimation error caused by statistical fluctuations of outcomes is called statistical error, and estimation error caused by systematic noise on the data is called systematic error. Usually systematic error is evaluated by assuming a type of systematic noise, for example, Gaussian noise. In this thesis we assume that we know the type of systematic noise and its effect is included in the random variables, so we can focus on the evaluation of statistical error. In statistical parameter estimation, there are at least two approaches for evaluating the amount of statistical error. One is called an expected loss, and the other is called an error probability. These are introduced in Sect. 3.1.3.
In order to define expected loss and error probability, we introduce a continuous two-variable function. A continuous two-variable function $\Delta : \mathbb{R}^k \times \mathbb{R}^k \to \mathbb{R}$ is called a loss function if it satisfies the following conditions for any $\theta, \theta', \theta'' \in \mathbb{R}^k$:

(i) (Non-negativity) $\Delta(\theta, \theta') \geq 0$.
(ii) $\Delta(\theta, \theta) = 0$.

We introduce three additional conditions:

(iii) (Symmetry) $\Delta(\theta, \theta') = \Delta(\theta', \theta)$.
(iv) (Triangle inequality) $\Delta(\theta, \theta') \leq \Delta(\theta, \theta'') + \Delta(\theta'', \theta')$.
(v) (Positivity) $\Delta(\theta, \theta') = 0 \Rightarrow \theta = \theta'$.

A loss function satisfying conditions (iii) and (iv) is called a semi-distance, and a semi-distance satisfying condition (v) is called a distance. A loss function satisfying condition (v) is called a pseudo-distance.

3.1.3 Figure of Merit

Suppose that we choose a loss function Δ. Then $\Delta(\theta_N^{\mathrm{est}}(D^N), \theta_0)$ is a random variable, and its value is given probabilistically. In this subsection, we introduce two approaches for removing the data-dependency of $\Delta(\theta_N^{\mathrm{est}}(D^N), \theta_0)$ (Table 3.5).

1. Expected loss
   Expected loss is a generalization of mean squared error. For an experimental design X^{exp}, an estimator θ^{est}, and a loss function Δ, the expected loss is defined as the expectation of the loss function between the estimate and true parameter, i.e.,

$$\bar{\Delta}_N(X^{\mathrm{exp}}, \theta^{\mathrm{est}}|\theta_0) := \mathbb{E}_{X^N,\theta_0}[\ \Delta(\theta_N^{\mathrm{est}}(D^N), \theta_0)\ ] \tag{3.14}$$

$$= \sum_{\mathscr{X}^N} p(\boldsymbol{x}^N|X^N, \theta_0)\ \Delta(\theta_N^{\mathrm{est}}(D^N), \theta_0). \tag{3.15}$$

   Expected loss is a statistical average of the loss between the estimate and true parameter, and is a generalization of error bars.

   Example 3.3 (Mean squared error)
   When we choose $\Delta(\theta, \theta') = \|\theta - \theta'\|^2$, the expected loss is equivalent to the mean squared error, $\mathbb{E}_{X^N,\theta_0}[\ \|\theta_N^{\mathrm{est}}(D^N) - \theta_0\|^2\ ]$.

2. Error probability
   Error probability is the probability that observation results deviate from the truth. Let δ be a positive number. We use δ as a threshold for statistical estimation error. Suppose that we consider an estimate as a failure when $\Delta(\theta_N^{\mathrm{est}}(D^N), \theta_0) > \delta$. For an experimental design X^{exp}, an estimator θ^{est}, a loss function Δ, and a positive value δ, the error probability is defined as the probability that we obtain deviate observation results, i.e.,

$$\mathrm{P}_{\delta,N}^{\Delta}(X^{\mathrm{exp}}, \theta^{\mathrm{est}}|\theta_0) := \mathbb{P}_{X^N,\theta_0}[\ \Delta(\theta_N^{\mathrm{est}}(D^N), \theta_0) > \delta\ ] \tag{3.16}$$

$$= \sum_{\boldsymbol{x}^N:\Delta(\theta_N^{\mathrm{est}}(D^N),\theta_0)>\delta} p(\boldsymbol{x}^N|X^N, \theta_0). \tag{3.17}$$

3. Average and maximum approaches
   An expected loss $\bar{\Delta}_N$ and error probability $\mathrm{P}_{\delta,N}^{\Delta}$ are functions for evaluating the estimation error of a given estimation scheme $(\mathscr{P}_\Theta, N, X^{\mathrm{exp}}, \theta^{\mathrm{est}})$ and a given true parameter θ_0. Theoretically we can analyze the behavior of $\bar{\Delta}_N$ and $\mathrm{P}_{\delta,N}^{\Delta}$ by assuming that the true parameter is θ_0, and this is called a pointwise analysis. However, we cannot directly apply the results of pointwise analysis to evaluating the estimation errors in experiments, because we do not know the true parameter

θ_0 in an experiment. We introduce two figures of merit that use expected loss or error probability and are applicable to evaluating estimation errors in experiments.

The first figure of merit is an average value of expected loss or error probability. Let μ denote a probability measure on Θ. The average expected loss and average error probability are defined as

$$\bar{\Delta}_N^{\mathrm{ave}}(X^{\mathrm{exp}}, \theta^{\mathrm{est}}) := \int_\Theta d\mu(\theta_0)\, \bar{\Delta}_N(X^{\mathrm{exp}}, \theta^{\mathrm{est}}|\theta_0), \tag{3.18}$$

$$\mathrm{P}_{\delta,N}^{\Delta\mathrm{ave}}(X^{\mathrm{exp}}, \theta^{\mathrm{est}}) := \int_\Theta d\mu(\theta_0)\, \mathrm{P}_{\delta,N}^{\Delta}(X^{\mathrm{exp}}, \theta^{\mathrm{est}}|\theta_0). \tag{3.19}$$

Average expected loss and average error probability are the average values of the expected loss and error probability over all possible true parameter θ_0 with weight μ, respectively. The values are understood as a "generic" expected loss and error probability in Θ.

The second figure of merit is the maximal value of expected loss or error probability.

$$\bar{\Delta}_N^{\max}(X^{\mathrm{exp}}, \theta^{\mathrm{est}}) := \operatorname*{argmax}_{\theta_0\in\Theta} \bar{\Delta}_N(X^{\mathrm{exp}}, \theta^{\mathrm{est}}|\theta_0), \tag{3.20}$$

$$\mathrm{P}_{\delta,N}^{\Delta\max}(X^{\mathrm{exp}}, \theta^{\mathrm{est}}) := \operatorname*{argmax}_{\theta_0\in\Theta} \mathrm{P}_{\delta,N}^{\Delta}(X^{\mathrm{exp}}, \theta^{\mathrm{est}}|\theta_0). \tag{3.21}$$

The maximal expected loss and error probability are the values of the expected loss and error probability in the worst case. By definition, for any $\theta_0 \in \Theta$ we have

$$\bar{\Delta}_N(X^{\mathrm{exp}}, \theta^{\mathrm{est}}|\theta_0) \le \bar{\Delta}_N^{\max}(X^{\mathrm{exp}}, \theta^{\mathrm{est}}), \tag{3.22}$$

$$\mathrm{P}_{\delta,N}^{\Delta}(X^{\mathrm{exp}}, \theta^{\mathrm{est}}|\theta_0) \le \mathrm{P}_{\delta,N}^{\Delta\max}(X^{\mathrm{exp}}, \theta^{\mathrm{est}}). \tag{3.23}$$

The choice of figure of merit depends on the purpose of the estimation experiment. In most quantum information situations, the purpose of the estimation experiment is not an estimation itself, but a verification of a successful quantum state or process implementation. For example, if the purpose of the experiment is to verify a successful preparation of a quantum state with rough accuracy, an average-type evaluation could be a candidate for the figure of merit. On the other hand, if we require a strict evaluation of estimation error, e.g., in the case that the experimentalists want to use the prepared state for a secure quantum cryptographic protocol and need to evaluate the degree of security rigorously, average-type evaluation is not suitable and maximum-type evaluation is applicable.

In this thesis, we analyze pointwise, maximum, and average loss functions in the test of Bell-type correlations and quantum tomography. In Sect. 4.3, we derive

Table 3.5 Notation regarding figures of merit

Notation	Explanation
$\Delta(\theta, \theta')$	A loss function
$\bar{\Delta}_N(X^{\mathrm{exp}}, \theta^{\mathrm{est}}\|\theta_0)$	A pointwise expected loss
$\bar{\Delta}_N^{\mathrm{ave}}(X^{\mathrm{exp}}, \theta^{\mathrm{est}})$	An average expected loss
$\bar{\Delta}_N^{\mathrm{max}}(X^{\mathrm{exp}}, \theta^{\mathrm{est}})$	A maximum expected loss
δ	A threshold of estimation error
$\mathrm{P}_{\delta,N}^{\Delta}(X^{\mathrm{exp}}, \theta^{\mathrm{est}}\|\theta_0)$	A pointwise error probability
$\mathrm{P}_{\delta,N}^{\Delta\mathrm{ave}}(X^{\mathrm{exp}}, \theta^{\mathrm{est}})$	An average error probability
$\mathrm{P}_{\delta,N}^{\Delta\mathrm{max}}(X^{\mathrm{exp}}, \theta^{\mathrm{est}})$	A maximum error probability

some functions which upper-bound maximum expected loss and maximum error probability. In Sect. 5.2, we derive such functions for maximum expected losses of linear and norm-minimization estimators, and analyze pointwise expected losses of a maximum-likelihood estimator. In Sect. 5.3, we derive some functions which upper-bound the maximum error probability of linear and norm-minimization estimators, and such functions for a maximum-likelihood estimator in a specific case. In Sect. 6.2, we analyze the pointwise and average expected losses in an adaptive quantum estimation.

3.2 Asymptotic Theory

In the previous section, we explained the fundamental concepts and technical terms in probability and statistical parameter estimation theory. In this section, we review known results from the asymptotic theory of statistical parameter estimation [1]. For a class of estimator, the behavior of expected loss and error probability around $N = \infty$ are explained. In Sect. 3.2.1, we explain some conditions assumed in asymptotic theory. In Sect. 3.2.2, we explain known results for expected losses in asymptotic theory. In Sect. 3.2.3, we explain known results for error probability in asymptotic theory. Known results for a specific estimator for finite N will be explained in Sect. 3.3.

3.2.1 Normal Conditions

Roughly speaking, as the number of measurement trials N becomes larger, more accurate parameter estimation is possible. For a given experimental design X^{exp} and estimator θ^{est}, we expect that the estimation error becomes smaller as N becomes larger, and that the estimation error converges to zero as N goes to infinity. For a given experimental design, an estimator is called consistent if the estimation error converges to zero as N goes to infinity for any true parameter. When we evaluate the estimation error by an expected loss using a loss function Δ, the consistent estimator is called Δ-consistent. When we evaluate by an error probability, the consistent estimator is called weakly consistent. Generally, sufficient conditions and necessary conditions for the existence of consistent estimators are very complicated. In this thesis, however, we treat only multinomial and Gaussian distributions, and in these cases the conditions are relatively simple.

First, we assume Condition 3.1, that is, the parameter space Θ is identifiable, as this is a necessary condition for the existence of consistent estimators. Next we impose a condition on the parameter space.

Condition 3.2 (Open parameter space)
Θ is a open set of $\mathbb{R}^k$.

When a parameter space is not open, the true parameter can be on a boundary of Θ. Standard techniques are then not applicable, and specialized analysis matching the case is necessary. The asymptotic behavior of estimation errors can be very different from the open parameter space case [2]. In order to avoid this difficulty, standard asymptotic theory requires Condition 3.2. In quantum tomography, however, the parameter space might not be open. We analyze such a case in Sect. 5.2.3. In this subsection however, we introduce asymptotic theory based on Condition 3.2.

Let ∇_θ denote the partial derivative vector, i.e., $\nabla_\theta^T := (\frac{\partial}{\partial\theta_1}, \ldots, \frac{\partial}{\partial\theta_k})$, where T denotes transposition. We assume differentiability of the probability distributions $p(\boldsymbol{x}^N | X^N, \theta)$ as many times as necessary. In the quantum case, this is satisfied. For a given parametrized probability distribution $\boldsymbol{p}_{X,\theta}$, let us define a $k \times k$ matrix $F(X, \theta)$ as

$$F(X, \theta) := \mathbb{E}_{X,\theta}[\, (\nabla_\theta \log p_{X,\theta})(\nabla_\theta^T \log p_{X,\theta})\,]. \tag{3.24}$$

$F(X, \theta)$ is called the Fisher information matrix of $\boldsymbol{p}_{X,\theta}$, or the Fisher matrix for short. When the parameter space is one-dimensional, i.e., $k = 1$, F is called the Fisher information.

Condition 3.3 (Inverse Fisher matrix)
For a given random variable sequence X^N, the Fisher matrix $F(X^N, \theta)$ is invertible, i.e., $F(X^N, \theta)^{-1}$ exists.

The inverse Fisher matrix plays an important role in the asymptotic theory of statistical parameter estimation. It characterizes the rate at which estimation errors decrease. The details are explained in Sects. 3.2.2 and 3.2.3.

3.2.2 *Expected Loss*

In this subsection we review the asymptotic theory of expected loss. Suppose that an estimation scheme $(\mathscr{P}_\Theta, N, X^{\mathrm{exp}}, \theta^{\mathrm{est}})$ and a loss function Δ are fixed.

Instead of treating general estimators, let us consider a limited class. An estimator θ^{est} is called asymptotically unbiased with respect to X^{exp} if

$$\lim_{N\to\infty} \nabla_\theta \mathbb{E}_{X^N,\theta}[\ \theta_N^{\mathrm{est}}(D^N)^T\ ] = I \tag{3.25}$$

holds for any $\theta \in \Theta$, where I is the identity matrix on $\mathbb{R}^k$.

Let us consider the covariance matrix of an estimator,

$$\mathbb{E}_{X^N,\theta_0}[\ (\theta_N^{\mathrm{est}} - \theta_0)(\theta_N^{\mathrm{est}} - \theta_0)^T\ ]. \tag{3.26}$$

For any asymptotically unbiased estimators, the following theorem holds.

Theorem 3.1 (Asymptotic Cramér-Rao inequality [3]) *Suppose that Conditions 3.1, 3.2, and 3.3 are satisfied. When an estimator θ^{est} is asymptotically unbiased, the inequality*

$$\lim_{N\to\infty} N \cdot \mathbb{E}_{X^N,\theta_0}[\ (\theta_N^{\mathrm{est}} - \theta_0)(\theta_N^{\mathrm{est}} - \theta_0)^T\ ] \geq \lim_{N\to\infty} N \cdot F(X^N, \theta_0)^{-1} \tag{3.27}$$

holds.

In many practical cases, $\mathbb{E}_{X^N,\theta_0}[\ (\theta_N^{\mathrm{est}} - \theta_0)^T(\theta_N^{\mathrm{est}} - \theta_0)\ ] = O(\frac{1}{N})$. The L.H.S. of Eq. (3.27) is the coefficient of the $\frac{1}{N}$ term. The R.H.S. gives a lower bound on the coefficient, and it is independent of the choice of the asymptotically unbiased estimator. Therefore the R.H.S. can be interpreted as a minimal estimation error for an experimental design for all asymptotically unbiased estimators. In particular, when the experimental design is i.i.d., the total Fisher matrix is given by sum of N Fisher matrices of a single random variable, i.e., $F(X^N, \theta) = N \cdot F(X, \theta)$, where $X_1 = \cdots = X_N = X$. The inverse is $F(X^N, \theta)^{-1} = F(X, \theta)^{-1}/N$, and we obtain

$$\lim_{N\to\infty} N \cdot \mathbb{E}_{\theta_0}[\ (\theta_N^{\mathrm{est}} - \theta_0)(\theta_N^{\mathrm{est}} - \theta_0)^T\ ] \geq F(X, \theta_0)^{-1}. \tag{3.28}$$

Equation (3.28) indicates that, for sufficiently large N, the covariance matrix of an asymptotically unbiased estimator can decrease at most as $O\left(\frac{1}{N}\right)$, and the minimum coefficients of decrease are given by the inverse Fisher matrix. A covariance matrix becomes smaller as the inverse Fisher matrix of an experimental design becomes smaller. This means that an experimental design with large Fisher matrix gives us more accurate estimation results. Under the same conditions, a maximum-likelihood estimator is asymptotically unbiased and attains the equality of Eq. (3.27). When the parameter is affine with respect to the probability distribution, i.e., $\theta = A\boldsymbol{p}_X + \boldsymbol{b}$, a linear estimator also attains the equality.

3.2.3 Error Probabilities

In this subsection we review the asymptotic theory of error probability. Suppose that an estimation scheme $(\mathscr{P}_\Theta, N, X^{\mathrm{exp}}, \theta^{\mathrm{est}})$ and a loss function Δ are fixed. When an estimator is weakly consistent, the error probability decreases exponentially as N goes to infinity, i.e.,

$$\mathrm{P}^{\Delta}_{\delta,N}(X^{\mathrm{exp}}, \theta^{\mathrm{est}}|\theta_0) \sim e^{-N \cdot r_{\mathrm{EP}}(\Delta,\delta,X^{\mathrm{exp}},\theta^{\mathrm{est}},\theta_0)}. \tag{3.29}$$

For a given Δ, δ, X^{exp}, and θ_0, the implicit form of the optimal bound on the rate of decrease r_{EP} is derived in [4]. The explicit forms are known in several specific cases [5–8]. For example, when the square of a loss function is a sufficiently smooth pseudo-distance with a non-zero Hesse matrix $\frac{1}{2}H(\theta)$, the explicit form of the optimal rate of decrease is given by [8]

$$r(\Delta, \delta, X^{\mathrm{exp}}, \theta) = \frac{\delta^2}{2\sigma_{\max}(\sqrt{H(\theta)}F(X,\theta)^{-1}\sqrt{H(\theta)})}, \tag{3.30}$$

where $\sigma_{\max}(A)$ denotes the maximal eigenvalue of a matrix A. As in the Cramér-Rao inequality, under the same conditions a maximum-likelihood estimator attains this optimal rate of decrease [8]. When the parameter is affine with respect to the probability distribution, a linear estimator also attains the optimal rate. From Eq. (3.30), a larger Fisher matrix gives a larger optimal rate of decrease.

3.3 Finite Theory for Sample Mean

In this section, we review known results for statistical errors on sample means for finite N. These results are used to analyze expected losses and error probabilities in the test of Bell-type correlations (Chap. 4) and in quantum tomography (Chap. 5).

Suppose that we observe a random variable sequence $X^N = \{X_1, \ldots, X_N\}$, and obtain an outcome sequence $\boldsymbol{x}^N = \{x_1, \ldots, x_N\}$. The sample sum $S_N(\boldsymbol{x}^N)$ and the sample mean $s_N(\boldsymbol{x}^N)$ are defined as

$$S_N(\boldsymbol{x}^N) := \sum_{n=1}^{N} x_n, \tag{3.31}$$

$$s_N(\boldsymbol{x}^N) := \frac{1}{N}\sum_{n=1}^{N} x_n. \tag{3.32}$$

When X^N is an i.i.d. random variable sequence, $X_1 = \cdots = X_N$ holds and the sample mean can be rewritten as

$$s_N(\boldsymbol{x}^N) = \sum_{x \in \mathscr{X}} \nu_{N,x}(\boldsymbol{x}^N)x, \tag{3.33}$$

where $\nu_{N,x}(\boldsymbol{x}^N)$ are the relative frequencies of $\boldsymbol{x}^N$ defined in Eq. (3.11).

3.3.1 Mean Squared Error

For a given probability distribution $\boldsymbol{p}_X$, we define a $M \times M$ matrix Γ by

$$\Gamma_{x,x'}(\boldsymbol{p}_X) := p_X(x)\delta_{x,x'} - p_X(x)p_X(x'), \tag{3.34}$$

From a straightforward calculation, we have the expectation and covariance matrix of relative frequencies as

$$\mathbb{E}_X[\boldsymbol{\nu}_N] = \boldsymbol{p}_X, \tag{3.35}$$

$$\mathbb{E}_X[(\boldsymbol{\nu}_N - \boldsymbol{p}_X)(\boldsymbol{\nu}_N - \boldsymbol{p}_X)^T] = \Gamma(\boldsymbol{p}_X). \tag{3.36}$$

Let us define an expectation with weight $\boldsymbol{a}$ as $A(\boldsymbol{p}) = \boldsymbol{a} \cdot \boldsymbol{p}_X = \sum_{x \in \mathscr{X}} a(x)p(x)$. Then $A(\boldsymbol{\nu}_N)$ is a linear estimate of $A(\boldsymbol{p}_X)$, and the mean squared error is given by

$$\mathbb{E}_X[|A(\boldsymbol{\nu}_N) - A(\boldsymbol{p}_X)|^2] = \frac{\boldsymbol{a}^T \Gamma(\boldsymbol{p}_X)\boldsymbol{a}}{N}. \tag{3.37}$$

This is the exact result for the linear estimator's expected loss. Equation (3.37) will be used for analyzing expected losses of linear estimators in test of Bell-type correlations (Sect. 4.3.2) and in quantum tomography (Sect. 5.2.1).

3.3.2 Tail Probability

First, we consider an independent random variable sequence X^N. Suppose that each X_n takes a value in $[a_n, b_n$. The following inequality holds [9].

$$\mathbb{P}_{X^N}[\ |S_N - \mathbb{E}[S_N]| > t\ ] \leq 2\exp\left[-2\frac{t^2}{\sum_{n=1}^N (b_n - a_n)^2}\right]. \tag{3.38}$$

Equation (3.38) is called Hoeffding's tail inequality, and gives an upper bound of the error probability of a sample sum. We will use Hoeffding's tail inequality for analyzing error probabilities in test of Bell-type correlations (Sect. 4.3.3) and in quantum tomography (Sect. 5.3.1).

Second, we consider an i.i.d. random variable sequence X^N. Let R denote a set of probability distributions with the sample space $\mathscr{X}$. The following inequality holds [10]

$$\mathbb{P}\left[\boldsymbol{\nu}_N \in R\right] \leq (N+1)^M \exp\left[-N \cdot \inf_{\boldsymbol{q} \in R} K(\boldsymbol{q} \| \boldsymbol{p}_X)\right], \tag{3.39}$$

where K is the Kullback-Libler divergence defined as

$$K(\boldsymbol{q} \| \boldsymbol{p}) := \sum_{x \in \mathscr{X}} q(x) \ln \frac{q(x)}{p(x)}. \tag{3.40}$$

Equation (3.39) is called Sanov's inequality. The exponential coefficient in Eq. (3.39), $(N+1)^M$, originates from over-counting a number of possible relative frequencies. By exactly counting the number, we can obtain an improved Sanov's inequality as follows:

$$\mathbb{P}\left[\boldsymbol{\nu}_N \in R\right] \leq \frac{(N+M-1)!}{N!(M-1)!} \exp\left[-N \cdot \inf_{\boldsymbol{q} \in R} K(\boldsymbol{q} \| \boldsymbol{p}_X)\right]. \tag{3.41}$$

We will use the improved Sanov's inequality for analyzing error probabilities for a maximum-likelihood estimator in quantum tomography (Sect. 5.3.3).

References

1. E.L. Lehmann, *Theory of Point Estimation*. Springer Texts in Statistics (Springer, 1998)
2. M. Akahira, K. Takeuchi, *Non-Regular Statistical Estimation*. Lecture Notes in Statistics (Springer, 1995)
3. C.R. Rao, *Linear Statistical Inference and Its Applications*, 2nd edn. Wiley Series in Probability and Statistics (Wiley, New York, 2002). (originally published in 1973)
4. R.R. Bahadur, J.C. Gupta, S.L. Zabell, *Large Deviations, Tests, and Estiamtes*. Asymptotic Theory of Statistical Tests and Estimates (Academic Press, New York, 1980), p. 33
5. R.R. Bahadur, Sankhyā **22**, 229 (1960)
6. R.R. Bahadur, Ann. Math. Statist. **38**, 303 (1967). doi:10.1214/aoms/1177698949
7. M. Hayashi, K. Matsumoto, IEICE Trans. **A 83**, 629 (2000)
8. T. Sugiyama, P.S. Turner, M. Murao, Phys. Rev. **A 83**, 012105 (2011). doi:10.1103/PhysRevA.83.012105
9. W. Hoeffding, J. Am. Stat. Assoc. **58**, 13 (1963). doi:10.2307/2282952
10. I.N. Sanov, (English translation from Mat. Sb. (42)) in. Selected Translations in Mathematical Statistics and Probability **1**(1961), (1957)

Chapter 4
Evaluation of Estimation Precision in Test of Bell-Type Correlations

4.1 Quantum Non-locality

First, we define product, separable and entangled states. Let us consider a composite system consisting of two quantum systems $\mathscr{H}_A$ and $\mathscr{H}_B$. The Hilbert space corresponding to the composite system is given by $\mathscr{H}_{AB} := \mathscr{H}_A \otimes \mathscr{H}_B$ (Postulate 3). Let us start with the case for pure states. A state vector $|\Psi\rangle_{AB} \in \mathscr{H}_{AB}$ is called a product state if it can be represented by a tensor product of two state vectors, $|\Psi\rangle_{AB} = |\psi\rangle_A \otimes |\varphi\rangle_B$. The corresponding density matrix is $|\Psi\rangle_{AB}\langle\Psi| = |\psi\rangle_A\langle\psi| \otimes |\varphi\rangle_B\langle\varphi|$. A state vector of a bipartite system is called an entangled state if it is not a product state. For example, the state defined by

$$|\Phi^+\rangle_{AB} := \frac{1}{\sqrt{2}}(|+1_z\rangle_A \otimes |+1_z\rangle_B + |-1_z\rangle_A \otimes |-1_z\rangle_B) \tag{4.1}$$

is an entangled state, where $|+1_z\rangle$ and $|-1_z\rangle$ are orthogonal to each other. An entangled state in the density matrix representation is defined similarly to the case in the state vector representation. A density matrix $\hat{\rho}_{AB} \in \mathscr{S}(\mathscr{H}_{AB})$ is called a separable state if it can be represented as a probabilistic mixture of product states, e.g., $\hat{\rho}_{AB} = \sum_i p_i |\psi_i\rangle_A\langle\psi_i| \otimes |\varphi_i\rangle_B\langle\varphi_i|$. A density matrix in a bipartite system is called an entangled state if it is not a separable state.

Next, we demonstrate a difference between product states and entangled states in terms of correlations of measurement outcomes. Suppose that we perform a measurement described by a POVM $\boldsymbol{\Pi}_{z,A} = \{|+1_z\rangle_A\langle+1_z|, |-1_z\rangle_A\langle-1_z|\}$ on system A, and perform a measurement $\boldsymbol{\Pi}_{z,B} = \{|+1_z\rangle_B\langle+1_z|, |-1_z\rangle_B\langle-1_z|\}$ on system B. The POVMs $\boldsymbol{\Pi}_{z,A}$ and $\boldsymbol{\Pi}_{z,B}$ correspond to the projective measurement of the observable

$$\hat{\sigma}_z := |+1_z\rangle\langle+1_z| - |-1_z\rangle\langle-1_z| \tag{4.2}$$

T. Sugiyama, *Finite Sample Analysis in Quantum Estimation*, Springer Theses, DOI: 10.1007/978-4-431-54777-8_4,

on A and B, respectively. When a state of the composite system $\hat{\rho}_{\mathrm{AB}}$ is given by $|\Phi^{+}\rangle_{\mathrm{AB}}\langle\Phi^{+}|$, using Eq. (2.2), the joint probability distribution $p(x_{\mathrm{A}}, x_{\mathrm{B}}|\boldsymbol{\Pi}_{\mathrm{z,A}} \otimes \boldsymbol{\Pi}_{\mathrm{z,B}}, \hat{\rho}_{\mathrm{AB}})$ is given by

$$\begin{aligned} p(+1_{\mathrm{z}}, +1_{\mathrm{z}}|\boldsymbol{\Pi}_{\mathrm{z,A}} \otimes \boldsymbol{\Pi}_{\mathrm{z,B}}, \hat{\rho}_{\mathrm{AB}}) &= 1/2, \\ p(+1_{\mathrm{z}}, -1_{\mathrm{z}}|\boldsymbol{\Pi}_{\mathrm{z,A}} \otimes \boldsymbol{\Pi}_{\mathrm{z,B}}, \hat{\rho}_{\mathrm{AB}}) &= 0, \\ p(-1_{\mathrm{z}}, +1_{\mathrm{z}}|\boldsymbol{\Pi}_{\mathrm{z,A}} \otimes \boldsymbol{\Pi}_{\mathrm{z,B}}, \hat{\rho}_{\mathrm{AB}}) &= 0, \\ p(-1_{\mathrm{z}}, -1_{\mathrm{z}}|\boldsymbol{\Pi}_{\mathrm{z,A}} \otimes \boldsymbol{\Pi}_{\mathrm{z,B}}, \hat{\rho}_{\mathrm{AB}}) &= 1/2. \end{aligned} \tag{4.3}$$

Therefore the probability of observing the same outcomes $x_{\mathrm{A}} = x_{\mathrm{B}}$ is one, and the probability of observing different outcomes $x_{\mathrm{A}} \neq x_{\mathrm{B}}$ is zero. This means that the measurement outcomes are perfectly correlated. There are separable states which also have perfect correlation between outcomes of $\boldsymbol{\Pi}_{\mathrm{z,A}}$ and $\boldsymbol{\Pi}_{\mathrm{z,B}}$. The state

$$\hat{\rho}_{\mathrm{AB}} = \frac{1}{2}|+1_{\mathrm{z}}\rangle_{\mathrm{A}}\langle +1_{\mathrm{z}}| \otimes |+1_{\mathrm{z}}\rangle_{\mathrm{B}}\langle +1_{\mathrm{z}}| + \frac{1}{2}|-1_{\mathrm{z}}\rangle_{\mathrm{A}}\langle -1_{\mathrm{z}}| \otimes |-1_{\mathrm{z}}\rangle_{\mathrm{B}}\langle -1_{\mathrm{z}}| \tag{4.4}$$

gives a joint probability distribution that is the same as Eq. (4.3). Thus we see that for the joint measurement $\boldsymbol{\Pi}_{\mathrm{z,A}} \otimes \boldsymbol{\Pi}_{\mathrm{z,B}}$, the entangled state and the product state give the same joint probability distribution. However, if we change the measurement, their joint probability distributions can vary. Let us introduce the notation

$$|+1_{\mathrm{x}}\rangle := \frac{1}{\sqrt{2}}(|+1_{\mathrm{z}}\rangle + |-1_{\mathrm{z}}\rangle), \tag{4.5}$$

$$|-1_{\mathrm{x}}\rangle := \frac{1}{\sqrt{2}}(|+1_{\mathrm{z}}\rangle - |-1_{\mathrm{z}}\rangle). \tag{4.6}$$

Suppose that we perform a measurement of $\boldsymbol{\Pi}_{\mathrm{x,A}} := \{|+1_{\mathrm{x}}\rangle_{\mathrm{A}}\langle +1_{\mathrm{x}}|, |-1_{\mathrm{x}}\rangle_{\mathrm{A}}\langle -1_{\mathrm{x}}|\}$ on system A and perform a measurement of $\boldsymbol{\Pi}_{\mathrm{x,B}} := \{|+1_{\mathrm{x}}\rangle_{\mathrm{B}}\langle +1_{\mathrm{x}}|, |-1_{\mathrm{x}}\rangle_{\mathrm{B}}\langle -1_{\mathrm{x}}|\}$ on system B. The POVMs $\boldsymbol{\Pi}_{\mathrm{x,A}}$ and $\boldsymbol{\Pi}_{\mathrm{x,B}}$ correspond to the projective measurement of the observable

$$\hat{\sigma}_{\mathrm{x}} := |+1_{\mathrm{x}}\rangle\langle +1_{\mathrm{x}}| - |-1_{\mathrm{x}}\rangle\langle -1_{\mathrm{x}}| \tag{4.7}$$

$$= |+1_{\mathrm{z}}\rangle\langle -1_{\mathrm{z}}| + |-1_{\mathrm{z}}\rangle\langle +1_{\mathrm{z}}| \tag{4.8}$$

on A and B, respectively. When the state of the composite system is the entangled state $|\Phi^{+}\rangle_{\mathrm{AB}}\langle\Phi^{+}|$, the joint distribution is given by

$$\begin{aligned} p(+1_{\mathrm{x}}, +1_{\mathrm{x}}|\boldsymbol{\Pi}_{\mathrm{x,A}} \otimes \boldsymbol{\Pi}_{\mathrm{x,B}}, \hat{\rho}_{\mathrm{AB}}) &= 1/2, \\ p(+1_{\mathrm{x}}, -1_{\mathrm{x}}|\boldsymbol{\Pi}_{\mathrm{x,A}} \otimes \boldsymbol{\Pi}_{\mathrm{x,B}}, \hat{\rho}_{\mathrm{AB}}) &= 0, \\ p(-1_{\mathrm{x}}, +1_{\mathrm{x}}|\boldsymbol{\Pi}_{\mathrm{x,A}} \otimes \boldsymbol{\Pi}_{\mathrm{x,B}}, \hat{\rho}_{\mathrm{AB}}) &= 0, \\ p(-1_{\mathrm{x}}, -1_{\mathrm{x}}|\boldsymbol{\Pi}_{\mathrm{x,A}} \otimes \boldsymbol{\Pi}_{\mathrm{x,B}}, \hat{\rho}_{\mathrm{AB}}) &= 1/2. \end{aligned} \tag{4.9}$$

As in the case of the measurement $\boldsymbol{\Pi}_{\mathrm{z,A}} \otimes \boldsymbol{\Pi}_{\mathrm{z,B}}$, the measurement outcomes are perfectly correlated. On the other hand, when the state is the separable state given by Eq. (4.4), the joint probability distribution is

$$\begin{aligned} p(+1_X, +1_X|\boldsymbol{\Pi}_{X,A} \otimes \boldsymbol{\Pi}_{X,B}, \hat{\rho}_{AB}) &= 1/4, \\ p(+1_X, -1_X|\boldsymbol{\Pi}_{X,A} \otimes \boldsymbol{\Pi}_{X,B}, \hat{\rho}_{AB}) &= 1/4, \\ p(-1_X, +1_X|\boldsymbol{\Pi}_{X,A} \otimes \boldsymbol{\Pi}_{X,B}, \hat{\rho}_{AB}) &= 1/4, \\ p(-1_X, -1_X|\boldsymbol{\Pi}_{X,A} \otimes \boldsymbol{\Pi}_{X,B}, \hat{\rho}_{AB}) &= 1/4, \end{aligned} \tag{4.10}$$

and the outcomes are completely uncorrelated.

To summarize this section, there are cases in which an entangled state exhibits a correlation between measurement outcomes whereas separable states do not. This implies that multiple joint probability distributions are necessary for testing quantum entanglement.

4.2 CHSH Inequality

As shown in the previous section, entangled states have a specific property that separable states do not. In this section, we explain the CHSH inequality that is used for testing whether a given state has such a property. The inequality is described in terms of probability distributions.

Suppose that there are two spatially separated systems labeled A and B. These systems are not necessarily quantum. There are two players, Alice and Bob. At one trial, each player independently flips a coin and obtains 0 for "heads" or 1 for "tails" with probability $1/2$. The pairs of possible results are $(0, 0)$, $(0, 1)$, $(1, 0)$, and $(1, 1)$, and their probabilities are equally $1/4$. Let i and j denote the flip result of Alice and Bob, respectively $(i, j = 0, 1)$. After the coin flip, Alice observes a random variables A_i, and Bob observes a random variable B_j. All four random variables, A_0, A_1, B_0, and B_1, take for their value either $+1$ or -1. Let $p(x_A, x_B|A_i, B_j)$ denote the joint probability distribution conditional on a coin flip result (i, j). The conditional joint probability distributions are explicitly given by

$$\begin{bmatrix} p(+1,+1|A_0,B_0) \\ p(+1,-1|A_0,B_0) \\ p(-1,+1|A_0,B_0) \\ p(-1,-1|A_0,B_0) \end{bmatrix}, \begin{bmatrix} p(+1,+1|A_0,B_1) \\ p(+1,-1|A_0,B_1) \\ p(-1,+1|A_0,B_1) \\ p(-1,-1|A_0,B_1) \end{bmatrix}, \begin{bmatrix} p(+1,+1|A_1,B_0) \\ p(+1,-1|A_1,B_0) \\ p(-1,+1|A_1,B_0) \\ p(-1,-1|A_1,B_0) \end{bmatrix}, \begin{bmatrix} p(+1,+1|A_1,B_1) \\ p(+1,-1|A_1,B_1) \\ p(-1,+1|A_1,B_1) \\ p(-1,-1|A_1,B_1) \end{bmatrix}. \tag{4.11}$$

We define a two-variable function η as

$$\eta(i, j) = \begin{cases} +1 \text{ if } (i, j) = (0, 0), (0, 1), (1, 0) \\ -1 \text{ if } (i, j) = (1, 1) \end{cases}. \tag{4.12}$$

Let us introduce an unconditional joint probability distribution $\boldsymbol{p} = \{p(i, j, x_A, x_B)\}_{(i,j,x_A,x_B)} \in \mathbb{R}^{16}$ as

$$p(i, j, x_A, x_B) = \frac{1}{4} p(x_A, x_B | A_i, B_j). \tag{4.13}$$

For a given $\boldsymbol{p}$, we define the CHSH correlation

$$C_{CHSH}(\boldsymbol{p}) := \mathbb{E}[A_0 \cdot B_0] + \mathbb{E}[A_0 \cdot B_1] + \mathbb{E}[A_1 \cdot B_0] - \mathbb{E}[A_1 \cdot B_1] \tag{4.14}$$

$$= \sum_{i,j} \sum_{x_A, x_B} p(x_A, x_B | A_i, B_j) \eta(i, j) \cdot x_A \cdot x_B \tag{4.15}$$

$$= \sum_{i,j} \sum_{x_A, x_B} p(i, j, x_A, x_B)\, 4\eta(i, j) \cdot x_A \cdot x_B, \tag{4.16}$$

where $\mathbb{E}[X_A \cdot X_B] = \sum_{x_A, x_B} p(x_A, x_B | X_A, X_B) x_A \cdot x_B$ is the expectation of the product of random variables X_A and X_B. The expectation takes values from $+1$ to -1 because the product of the outcomes $x_A \cdot x_B$ is $+1$ or -1. Therefore we have

$$-4 \leq C_{CHSH}(\boldsymbol{p}) \leq +4. \tag{4.17}$$

Noe we introduce a local hidden-variable model. Let $x_{A,ij}$ and $x_{B,ij}$ denote the outcomes of X_A and X_B when $X_A = A_i$ and $X_B = B_j$ $(i, j = 0, 1)$. The CHSH correlation is rewritten as

$$C_{CHSH}(\boldsymbol{p}) = \sum_{\substack{(x_{A,00}, x_{B,00}), (x_{A,01}, x_{B,01}), \\ (x_{A,10}, x_{B,10}), (x_{A,11}, x_{B,11})}} \left\{ \prod_{i,j} p(x_{A,ij}, x_{B,ij} | A_i, B_j) \right\} \times \left(x_{A,00} \cdot x_{B,00} + x_{A,01} \cdot x_{B,01} + x_{A,10} \cdot x_{B,10} - x_{A,11} \cdot x_{B,11} \right). \tag{4.18}$$

When two random variables X_A and X_B are independent, the joint distribution is called local. Roughly speaking, this means that the value of Alice's random variable X_A is independent of the choice of Bob's random variable X_B, and vice versa. Let $\boldsymbol{\lambda}$ be a vector in an arbitrary-dimensional Euclidean space, and f_X be a function labeled with a random variable X from the Euclidean space to $\{0, 1\}$. Suppose that $\boldsymbol{\lambda}$ is given with probability density $\mu(\boldsymbol{\lambda})$. When a probability distribution $\{p(x|X)\}_{x \in \mathscr{X}}$ can be represented as

$$p(x|X) = \int d\mu(\boldsymbol{\lambda}) p(x|X, \boldsymbol{\lambda}), \tag{4.19}$$

$$p(x|X, \boldsymbol{\lambda}) = \delta_{x, f_X(\boldsymbol{\lambda})}, \tag{4.20}$$

the probability distribution is called a hidden-variable model. The vector $\boldsymbol{\lambda}$ is called the hidden variable. Equation (4.20) means that, when we observe a random variable X, for a given $\boldsymbol{\lambda}$, an outcome $x = f_X(\boldsymbol{\lambda})$ is obtained with probability one. Equation (4.19) means that, when we observe random variables X, for a given $\boldsymbol{\lambda}$, the outcome $x = f_X(\boldsymbol{\lambda})$ is obtained deterministically, but because of our lack of knowledge of the value of $\boldsymbol{\lambda}$, the outcomes behave probabilistically. This is a classical interpretation of the origin of probabilistic phenomena, and corresponds to classical mechanics. When a set of joint probability distributions is a hidden-variable model and the hidden-variable model is local, i.e.,

$$p(x_{\mathrm{A}}, x_{\mathrm{B}}|A_i, B_j) = \int d\mu(\boldsymbol{\lambda}) p(x_{\mathrm{A}}|A_i, \boldsymbol{\lambda}) p(x_{\mathrm{B}}|B_j, \boldsymbol{\lambda}), \tag{4.21}$$

$$p(x_{\mathrm{A}}|A_i, \boldsymbol{\lambda}) = \delta_{x_{\mathrm{A}}, f_{A_i}(\boldsymbol{\lambda})}, \tag{4.22}$$

$$p(x_{\mathrm{B}}|B_j, \boldsymbol{\lambda}) = \delta_{x_{\mathrm{B}}, f_{B_j}(\boldsymbol{\lambda})}, \tag{4.23}$$

it is called a local hidden-variable model. For a local hidden-variable model, the following theorem holds.

Theorem 4.1 (CHSH inequality [1])
When four joint probability distributions are described by a local hidden-variable model, the value of the CHSH correlation is restricted to be between +2 and −2, i.e.,

$$|C_{\mathrm{CHSH}}(\boldsymbol{p})| \leq 2. \tag{4.24}$$

Proof (Theorem 4.1)
When a set of joint probability distributions obeys a local hidden-variable model, for a given $\boldsymbol{\lambda}$, the outcome of each player's random variable is independent of the choice of the other player's random variable. Then we have $x_{\mathrm{A},ij} = a_i(\boldsymbol{\lambda})$, $x_{\mathrm{B},ij} = b_j(\boldsymbol{\lambda})$, and

$$\begin{aligned} & x_{\mathrm{A},00} \cdot x_{\mathrm{B},00} + x_{\mathrm{A},01} \cdot x_{\mathrm{B},01} + x_{\mathrm{A},10} \cdot x_{\mathrm{B},10} - x_{\mathrm{A},11} \cdot x_{\mathrm{B},11} \\ & = a_0(\boldsymbol{\lambda}) \cdot b_0(\boldsymbol{\lambda}) + a_0(\boldsymbol{\lambda}) \cdot b_1(\boldsymbol{\lambda}) + a_1(\boldsymbol{\lambda}) \cdot b_0(\boldsymbol{\lambda}) - a_1(\boldsymbol{\lambda}) \cdot b_1(\boldsymbol{\lambda}). \end{aligned} \tag{4.25}$$

By substituting ±1 into $a_0(\boldsymbol{\lambda}), a_1(\boldsymbol{\lambda}), b_0(\boldsymbol{\lambda}), b_1(\boldsymbol{\lambda})$ in the R.H.S. of Eq. (4.25), we have

$$-2 \leq a_0(\boldsymbol{\lambda}) \cdot b_0(\boldsymbol{\lambda}) + a_0(\boldsymbol{\lambda}) \cdot b_1(\boldsymbol{\lambda}) + a_1(\boldsymbol{\lambda}) \cdot b_0(\boldsymbol{\lambda}) - a_1(\boldsymbol{\lambda}) \cdot b_1(\boldsymbol{\lambda}) \leq +2. \tag{4.26}$$

Therefore by integrating Eq. (4.26) over $\boldsymbol{\lambda}$, we obtain

$$-2 \leq C_{\mathrm{CHSH}}(\boldsymbol{p}) \leq +2. \tag{4.27}$$

□

Eq. (4.24) is called the CHSH inequality.

As shown in Eq. (4.17), the CHSH correlation can be larger than 2 or smaller than −2. If the CHSH correlation takes a value larger than 2 or smaller −2, it means that no local hidden-variable models can describe the joint probability distributions. This is called a violation of the CHSH inequality. It is known that the CHSH inequality can be violated in quantum mechanics.

Theorem 4.2 (Tsirelson's inequality [2])
When the joint probability distributions obey quantum mechanics, i.e.,

$$p(x_{\mathrm{A}}, x_{\mathrm{B}}|X_{\mathrm{A}}, X_{\mathrm{B}}) = \mathrm{Tr}[(\hat{\Pi}_{\mathrm{A},x_{\mathrm{A}}} \otimes \hat{\Pi}_{\mathrm{B},x_{\mathrm{B}}})\hat{\rho}_{\mathrm{AB}}] \tag{4.28}$$

hold, we have

$$-2\sqrt{2} \leq C_{\mathrm{CHSH}}(\boldsymbol{p}) \leq 2\sqrt{2}. \tag{4.29}$$

Equation (4.29) is called Tsirelson's inequality, and the upper and lower bounds are called Tsirelson's bounds. We note that Eq. (4.29) is valid for any bipartite quantum systems. We omit the proof but show that Tsirelson's bound is attainable in the case of $\mathscr{H}_{\mathrm{A}} = \mathscr{H}_{\mathrm{B}} = \mathbb{C}^2$. Suppose that the state of the composite system is described by $|\Phi^+\rangle_{\mathrm{AB}}$ defined by Eq. (4.1). Let $\boldsymbol{\Pi}_{\mathrm{A}0}, \boldsymbol{\Pi}_{\mathrm{A}1}, \boldsymbol{\Pi}_{\mathrm{B}0}$, and $\boldsymbol{\Pi}_{\mathrm{B}1}$ denote the POVMs corresponding to the random variables A_0, A_1, and B_0, B_1, respectively. We define two matrices $\hat{\sigma}_+$ and $\hat{\sigma}_-$ as

$$\hat{\sigma}_+ := \frac{1}{\sqrt{2}}(\hat{\sigma}_{\mathrm{z}} + \hat{\sigma}_{\mathrm{x}}) \tag{4.30}$$

$$\hat{\sigma}_- := \frac{1}{\sqrt{2}}(\hat{\sigma}_{\mathrm{z}} - \hat{\sigma}_{\mathrm{x}}). \tag{4.31}$$

We choose the projective measurements of $\hat{\sigma}_{\mathrm{z}}$, $\hat{\sigma}_{\mathrm{x}}$, $\hat{\sigma}_+$, and $\hat{\sigma}_-$ as $\boldsymbol{\Pi}_{\mathrm{A}0}, \boldsymbol{\Pi}_{\mathrm{A}1}, \boldsymbol{\Pi}_{\mathrm{B}0}$, and $\boldsymbol{\Pi}_{\mathrm{B}1}$. Then we obtain

$$\mathbb{E}[A_0 \cdot B_0] = \mathrm{Tr}[(\hat{\sigma}_{\mathrm{z}} \otimes \hat{\sigma}_+)|\Phi^+\rangle_{\mathrm{AB}}\langle\Phi^+|] = +\frac{1}{\sqrt{2}}, \tag{4.32}$$

$$\mathbb{E}[A_0 \cdot B_1] = \mathrm{Tr}[(\hat{\sigma}_{\mathrm{z}} \otimes \hat{\sigma}_-)|\Phi^+\rangle_{\mathrm{AB}}\langle\Phi^+|] = +\frac{1}{\sqrt{2}}, \tag{4.33}$$

$$\mathbb{E}[A_1 \cdot B_0] = \mathrm{Tr}[(\hat{\sigma}_{\mathrm{x}} \otimes \hat{\sigma}_+)|\Phi^+\rangle_{\mathrm{AB}}\langle\Phi^+|] = +\frac{1}{\sqrt{2}}, \tag{4.34}$$

$$\mathbb{E}[A_1 \cdot B_1] = \mathrm{Tr}[(\hat{\sigma}_{\mathrm{x}} \otimes \hat{\sigma}_-)|\Phi^+\rangle_{\mathrm{AB}}\langle\Phi^+|] = -\frac{1}{\sqrt{2}}. \tag{4.35}$$

Therefore in this case the CHSH correlation takes $2\sqrt{2}$ and Tsirelson's bound is attained.

4.3 Test of the CHSH Inequality

A violation of the CHSH inequality indicates that there exists a correlation between systems A and B that cannot be described by any local hidden-variable model. This does not imply the existence of entanglement in the systems because the CHSH inequality is described in terms of probability theory and the joint distributions do not necessarily obey quantum mechanics. Only after an experiment verifies that the bipartite system obeys quantum mechanics and the violation of the CHSH inequality is observed, can we insist that quantum entanglement is present.

4.3.1 Estimation Setting

As shown in Sect. 4.2, the purpose of the CHSH experiment is to find the value of the CHSH correlation of a given set of joint probability distributions. Our estimation object is the CHSH correlation defined by Eq. (4.14). At each trial, we choose a pair of random variables from the set $\{(A_0, B_0), (A_0, B_1), (A_1, B_0), (A_1, B_1)\}$ with equal probability, $1/4$. Let I_N and J_N denote random variables for the choice of the pair at the N-th trial, and i_N and j_N be outcomes of I_N and J_N, respectively. For a given pair (i_N, j_N), the pair of random variables chosen is (A_{i_N}, B_{j_N}), $(i_N, j_N = 0, 1)$. In the experimental design for testing the CHSH inequality, the N-th random variable X_N is defined by $X_N := (I_N, J_N, A_{I_N}, B_{J_N})$. The probability distribution is given by

$$p(i, j, x_A, x_B) = \frac{1}{4} \cdot p(x_A, x_B | A_i, B_j). \tag{4.36}$$

For a given data sequence $\boldsymbol{x}^N$, let $\boldsymbol{\nu}_N(\boldsymbol{x}^N) = \{\nu_{N,(i,j,x_A,x_B)}\}_{(i,j,x_A,x_B)}$ denote the relative frequency. We choose the CHSH correlation of $\boldsymbol{\nu}_N(\boldsymbol{x}^N)$ as the estimate. This is a linear estimator we will denote by C_N^L,

$$C_N^L(\boldsymbol{x}^N) := C_{CHSH}(\boldsymbol{\nu}_N(\boldsymbol{x}^N)) \tag{4.37}$$

$$= \sum_{i,j,x_A,x_B} \nu_{N,(i,j,x_A,x_B)}(\boldsymbol{x}^N)\, 4\eta(i, j) \cdot x_A \cdot x_B. \tag{4.38}$$

From the law of large numbers, we have

$$C_N^L(\boldsymbol{x}^N) \underset{N\to\infty}{\to} \sum_{i,j,x_A,x_B} p(i, j, x_A, x_B)\, 4\eta(i, j) \cdot x_A \cdot x_B \tag{4.39}$$

$$= \sum_{i,j,x_A,x_B} p(x_A, x_B | A_i, B_j)\eta(i, j) \cdot x_A \cdot x_B \tag{4.40}$$

$$= C_{CHSH}(\boldsymbol{p}). \tag{4.41}$$

Therefore the estimate C_N^L converges to the true value as N goes to infinity.

The estimation object is a real value, and we choose either the absolute value of the difference between the estimate and the true value or its square as the loss function, i.e.,

$$\Delta^{\mathrm{abs}}(C_N^{\mathrm{est}}, C_{\mathrm{CHSH}}(\boldsymbol{p})) = |C_N^{\mathrm{est}} - C_{\mathrm{CHSH}}(\boldsymbol{p})|, \tag{4.42}$$
$$\Delta^{\mathrm{abs2}}(C_N^{\mathrm{est}}, C_{\mathrm{CHSH}}(\boldsymbol{p})) = |C_N^{\mathrm{est}} - C_{\mathrm{CHSH}}(\boldsymbol{p})|^2. \tag{4.43}$$

4.3.2 Expected Loss

In this subsection, we analyze the expected loss of the loss function Δ^{abs2}. This is the mean squared error. We have the following theorem.

Theorem 4.3 (Expected loss, C^{L})
Let $\boldsymbol{p} = \{p(i, j, x_{\mathrm{A}}, x_{\mathrm{B}})\}_{(i,j,x_{\mathrm{A}},x_{\mathrm{B}})}$ denote the true probability distribution. The mean squared error of C^{L} and its upper bound are given by

$$\bar{\Delta}_N^{\mathrm{abs2}}(X^{\mathrm{CHSH}}, C^{\mathrm{L}}|\boldsymbol{p}) = \frac{16 - |C_{\mathrm{CHSH}}(\boldsymbol{p})|^2}{N} \tag{4.44}$$
$$\leq \frac{16}{N}. \tag{4.45}$$

The equality in Eq. (4.45) is attained by uniform distribution.

Proof (Theorem 4.3)
By substituting Eqs. (4.16) and (4.38) into Eq. (3.37), we obtain

$$\bar{\Delta}_N^{\mathrm{abs2}}(X^{\mathrm{CHSH}}, C^{\mathrm{L}}|\boldsymbol{p}) = \frac{\boldsymbol{v}^T \Gamma(\boldsymbol{p}) \boldsymbol{v}}{N}, \tag{4.46}$$

where $\boldsymbol{v}$ is a Euclidean vector in $\mathbb{R}^{16}$ defined by

$$v_{i,j,x_{\mathrm{A}},x_{\mathrm{B}}} := 4\eta(i, j) \cdot x_{\mathrm{A}} \cdot x_{\mathrm{B}}. \tag{4.47}$$

By using Eq. (3.34) and the relation $\boldsymbol{v} \cdot \boldsymbol{p} = C_{\mathrm{CHSH}}(\boldsymbol{p})$, we obtain Eq. (4.44). Equation (4.45) is obtained immediately from Eqs. (4.44) and (4.17). □

From Eq. (4.44), we can see that the range of the expected loss changes as follows:

$$0 \leq \bar{\Delta}_N^{\mathrm{abs2}}(X^{\mathrm{CHSH}}, C^{\mathrm{L}}|\boldsymbol{p}) < \frac{8}{N}, \quad (2\sqrt{2} < |C_{\mathrm{CHSH}}(\boldsymbol{p})| \leq 4), \tag{4.48}$$
$$\frac{8}{N} \leq \bar{\Delta}_N^{\mathrm{abs2}}(X^{\mathrm{CHSH}}, C^{\mathrm{L}}|\boldsymbol{p}) < \frac{12}{N}, \quad (2 < |C_{\mathrm{CHSH}}(\boldsymbol{p})| \leq 2\sqrt{2}), \tag{4.49}$$
$$\frac{12}{N} \leq \bar{\Delta}_N^{\mathrm{abs2}}(X^{\mathrm{CHSH}}, C^{\mathrm{L}}|\boldsymbol{p}) \leq \frac{16}{N}, \quad (0 \leq |C_{\mathrm{CHSH}}(\boldsymbol{p})| \leq 2). \tag{4.50}$$

Equation (4.48) corresponds to the super-quantum joint probability distributions. Equation (4.49) corresponds to joint probability distributions that can be described by quantum mechanics but cannot be described by any local hidden-variable model. Equation (4.50) corresponds to those which can be described by local hidden-variable models.

4.3.3 *Error Probability*

In this subsection, we derive a function that upper-bounds the error probability by using the Hoeffding inequality. Almost same results for statistical hypothesis testing were obtained using same techniques in [3].

Theorem 4.4 (Error probability, C^{L}**)**
For any probability distribution $\boldsymbol{p} \in \mathbb{R}^{16}$,

$$\mathrm{P}^{\mathrm{abs}}_{\delta,N}(X^{\mathrm{CHSH}}, C^{\mathrm{L}}|\boldsymbol{p}) \leq 2\exp(-N\delta^2/32). \tag{4.51}$$

Proof (Theorem 4.4)
By substituting $\eta(I_N, J_N) \cdot X_{\mathrm{A}} \cdot X_{\mathrm{B}}$ into the random variable in Eq. (3.38) and $b_n - a_n = 4 - (-4) = 8$, we obtain Eq. (4.51). □

When we perform the CHSH experiment with N trials and obtain an estimate $C^{\mathrm{L}}_N(\boldsymbol{x}^N)$, the probability that the estimate is included in the region $[C_{\mathrm{CHSH}}(\boldsymbol{p}) - \delta, C_{\mathrm{CHSH}}(\boldsymbol{p}) + \delta]$ is at least as large as $1 - 2e^{-N\delta^2/32}$.

4.4 Summary

In this chapter, we analyzed the estimation errors in a test of the CHSH inequality with finite data. In Sect. 4.1, we explained non-locality in quantum mechanics as well as the difference between local hidden-variable models and quantum mechanics from the perspective of the correlations of measurement outcomes. In Sect. 4.2, we explained the CHSH inequality and its relationship with quantum entanglement. In Sect. 4.3, we analyzed the expected loss and error probability of the arithmetic mean in the CHSH experiment. Explicit formulae for the upper bounds of the expected loss and error probability were given. The mathematical techniques used in Sect. 4.3 are also applicable for the other tests using correlation functions that consist of a linear combination of joint probability distributions, for example, a linear entanglement witness [4] and the KCBS inequality [5].

References

1. J.F. Clauser, M.A. Horn, A. Shimony, R.A. Holt, Phys. Rev. Lett. **23**, 880 (1969). doi:10.1103/PhysRevLett.23.880
2. B.S. Cirel'son, Lett. Math. Phys. **4**, 93 (1980). doi:10.1007/BF00417500
3. R.D. Gill, arXiv:1207.5103 [stat.AP]
4. O. Gühne, G. Tóth, Phys. Rep. **474**, 1 (2009). doi:10.1016/j.physrep.2009.02.004
5. A.A. Klyachko, M.A. Can, S. Binicioğlu, A.S. Shumovsky, Phys. Rev. Lett. **101**, 020403 (2008). doi:10.1103/PhysRevLett.101.020403

Chapter 5
Evaluation of Estimation Precision in Quantum Tomography

5.1 Estimation Setting

In this section, we discuss the estimation setting in quantum tomography. In Sect. 5.1.1, we categorize quantum tomography into four types according to the estimation objects: quantum state tomography, quantum process tomography, POVM tomography, and quantum instrument tomography. In each type, we assume that a sequence of identical copies of the unknown estimation object is given, and we try to identify the estimation object from data obtained by quantum measurements performed on the copies. In Sect. 5.1.2, we explain the estimation setting in quantum tomography—for simplicity, we focus on the estimation setting in quantum state tomography.

5.1.1 Estimation Objects

First, we explain estimation objects in quantum tomography, and give details of their matrix representation and parametrization. This parametrization reduces quantum tomography into statistical parameter estimation.

Suppose that a quantum system is given and we know the corresponding Hilbert space $\mathscr{H}$. Let $d := \dim \mathscr{H}$ denote the dimension of the Hilbert space, and $\mathscr{S}(\mathscr{H})$ be the set of all density matrices acting on the Hilbert space. In quantum tomography, estimation objects fall into four categories: state, process, probability distribution of measurement outcomes, and state transition caused by measurements. As we explained in Sect. 2.1, the mathematical representations are: density matrix $\hat{\rho}$, linear, completely, positive and trace-preserving map κ, positive operator-valued measure $\boldsymbol{\Pi} = \{\hat{\Pi}_x\}_{x \in \mathscr{X}}$, and quantum instruments $\boldsymbol{\kappa} = \{\kappa_x\}_{x \in \mathscr{X}}$, respectively. We refer to types of quantum tomography for these estimation objects as quantum state tomography (QST) [1–3], quantum process tomography (QPT) [4–8], POVM tomography

T. Sugiyama, *Finite Sample Analysis in Quantum Estimation*, Springer Theses,
DOI: 10.1007/978-4-431-54777-8_5,

Table 5.1 Categories of estimation objects in quantum tomography

Type	Tomographic object	Mathematical representation	Parameter number
QST	Quantum state	Density matrix, *hatρ*	$d^2 - 1$
QPT	Quantum process	LCPTP map, κ	$d^4 - d^2$
POVMT	Quantum measurement (Probability distribution)	POVM, $\boldsymbol{\Pi}$	$(M-1)d^2$
QIT	Quantum measurement (State transition)	Quantum instrument, $\boldsymbol{\kappa}$	$Md^4 - d^2$

(POVMT) [9–11], and quantum instrument tomography (QIT) [11], respectively. Table 5.1 summarizes these categories. M is the number of elements in $\mathscr{X}$.

5.1.2 Quantum State Tomography

In this subsection, we explain the estimation setting in quantum state tomography. The objective of quantum state tomography is to identify the density matrix $\hat{\rho}$ describing the state of the system.

5.1.2.1 One-qubit Case

Let us start with a simple example, a two-level system. This system is called a qubit, and is a fundamental element in quantum information. This estimation problem is also referred to as one-qubit state tomography.

1. **Estimation object and parametrization**
   In one-qubit state tomography ($\mathscr{H} = \mathbb{C}^2,\ d = 2$), the estimation object is a 2×2 density matrix $\hat{\rho}$, i.e.,

$$\hat{\rho} = \begin{pmatrix} \rho_{11} & \rho_{12} \\ \rho_{21} & \rho_{22} \end{pmatrix}, \quad \rho_{\alpha\beta} \in \mathbb{C}\ (\alpha, \beta = 1, 2). \tag{5.1}$$

   From the Hermiticity and trace-1 property, $\hat{\rho}$ can be parametrized by three real parameters $\boldsymbol{s} = (s_1, s_2, s_3)^T$ [12]

$$\hat{\rho} = \hat{\rho}(\boldsymbol{s}) = \frac{1}{2}\left(\hat{\mathbb{1}} + \boldsymbol{s} \cdot \boldsymbol{\sigma}\right) \tag{5.2}$$

$$= \frac{1}{2}\begin{pmatrix} 1 + s_3 & s_1 - is_2 \\ s_1 + is_2 & 1 - s_3 \end{pmatrix}, \tag{5.3}$$

$$\boldsymbol{\sigma} = (\hat{\sigma}_1,\ \hat{\sigma}_2,\ \hat{\sigma}_3)^T, \tag{5.4}$$

   where $\hat{\mathbb{1}}$ is the identity matrix and $\hat{\sigma}_1,\ \hat{\sigma}_2$, and $\hat{\sigma}_3$ are the Pauli matrices

$$\hat{\sigma}_1 := \begin{pmatrix} 0 & 1 \\ 1 & 0 \end{pmatrix},\ \hat{\sigma}_2 := \begin{pmatrix} 0 & -i \\ i & 0 \end{pmatrix},\ \hat{\sigma}_3 := \begin{pmatrix} 1 & 0 \\ 0 & -1 \end{pmatrix}. \tag{5.5}$$

The vector $\boldsymbol{s}$ is an element of three-dimensional Euclidean space $\mathbb{R}^3$ and the positive semidefiniteness restricts the range to a unit ball, $\|\boldsymbol{s}\|_2 = \sqrt{(s_1)^2 + (s_2)^2 + (s_3)^2} \leq 1$. This is called a Bloch vector, and the set of Bloch vectors

$$B := \{\boldsymbol{s} \in \mathbb{R}^3 |\ \|\boldsymbol{s}\|_2 \leq 1\} \tag{5.6}$$

is called the Bloch sphere. Bloch vectors on the surface $\|\boldsymbol{s}\|_2 = 1$ correspond to pure states. For a given density matrix $\hat{\rho}$, the corresponding Bloch vector $\boldsymbol{s}$ can be calculated as

$$s_\alpha = \mathrm{Tr}[\hat{\rho}\hat{\sigma}_\alpha],\ \alpha = 1, 2, 3. \tag{5.7}$$

The R.H.S. of Eq. (5.2) is called the Bloch representation or the Bloch parametrization of the density matrix $\hat{\rho}$. This is a bijective parametrization of density matrices, and estimating the Bloch vector $\boldsymbol{s}$ is equivalent to estimating $\hat{\rho}$. When we obtain an estimate of the Bloch vector $\boldsymbol{s}_N^{\mathrm{est}}$, the corresponding estimate of the density matrix is given by $\hat{\rho}(\boldsymbol{s}_N^{\mathrm{est}}) = \frac{1}{2}(\hat{\mathbb{1}} + \boldsymbol{s}_N^{\mathrm{est}} \cdot \boldsymbol{\sigma})$ By introducing this parametrization, we can apply the theory of statistical parameter estimation for quantum state tomography.

2. **Experimental design**

Equation (5.7) indicates that for a given density matrix $\hat{\rho}$, the Bloch vector $\boldsymbol{s}$ is given as the expectations of Pauli measurements. This gives us an experimental design for estimating $\boldsymbol{s}$. Suppose that we perform a projective measurement of $\hat{\sigma}_3$. The POVM is $\boldsymbol{\Pi} = \{|3, +1\rangle\langle 3, +1|, |3, -1\rangle\langle 3, -1|\}$ ($+1$ and -1 are outcomes), where $|\alpha, \pm 1\rangle$ are the eigenvectors of $\hat{\sigma}_\alpha$ ($\alpha = 1, 2, 3$). The third element of the Bloch vector is calculated from the probability distribution as $s_3 = p(3, +1|\boldsymbol{\Pi}, \hat{\rho}) - p(3, -1|\boldsymbol{\Pi}, \hat{\rho})$. If we could perform this projective measurement infinitely many times, we could learn the probability distribution by counting ratios of outcomes, and then could calculate the exact value of s_3 from the probability distribution. In order to identify all three elements of the Bloch vector, we also need to perform the projective measurement of $\hat{\sigma}_1$ and $\hat{\sigma}_2$. Let us consider the case that we perform the projective measurement of each Pauli matrix with probability $\frac{1}{3}$. The corresponding POVM is given by

$$\boldsymbol{\Pi}^{(6)} := \left\{ \frac{1}{3}|\alpha, +1\rangle\langle\alpha, +1|, \frac{1}{3}|\alpha, -1\rangle\langle\alpha, -1| \right\}_{\alpha=1,2,3.}. \tag{5.8}$$

In this case, $(\alpha, +1)$ and $(\alpha, -1)$ are possible measurement outcomes ($\alpha = 1, 2, 3$). This POVM is called the six-state POVM and is used in many one-qubit state tomography experiments. We can calculate all three parameters from the probability distribution of $\boldsymbol{\Pi}^{(6)}$ by

$$s_1 = 3\left\{p(1, +1|\boldsymbol{\Pi}^{(6)}, \hat{\rho}) - p(1, -1|\boldsymbol{\Pi}^{(6)}, \hat{\rho})\right\}, \tag{5.9}$$

$$s_2 = 3\left\{p(2, +1|\boldsymbol{\Pi}^{(6)}, \hat{\rho}) - p(2, -1|\boldsymbol{\Pi}^{(6)}, \hat{\rho})\right\}, \tag{5.10}$$

$$s_3 = 3\left\{p(3, +1|\boldsymbol{\Pi}^{(6)}, \hat{\rho}) - p(3, -1|\boldsymbol{\Pi}^{(6)}, \hat{\rho})\right\}. \tag{5.11}$$

3. **Estimator**

In one-qubit state tomography, we assume that a sequence of identical copies of the unknown state of a qubit is given, and we try to identify the Bloch vector describing the state from data obtained by quantum measurements performed on the copies. As explained in the previous paragraph, if infinitely many copies are given and we can perform quantum measurements infinitely many times, we can completely identify the Bloch vector, for example, by using Eqs. (5.9), (5.10) and (5.11). In real experiments, however, the number of copies are finite, and we cannot perform measurements infinitely many times. We need to choose a data-processing method for estimating the Bloch vector from a finite measurement sample. Such a method is an estimator in one-qubit state tomography.

Suppose that we performed the quantum measurement described by the six-state POVM N times and obtained a sequence of measurement outcomes $\boldsymbol{x}^N = \{x_1, \ldots, x_N\}$. Let $\boldsymbol{\nu}_N = \{\nu_N(\alpha, +1), \nu_N(\alpha, -1)\}_{\alpha=1,2,3}$ denote the relative frequencies of outcomes with respect to $\boldsymbol{x}^N$. When N is sufficiently large, the law of large number guarantees that the relative frequencies converge to the true probability distribution. It seems to be natural to substitute the relative frequencies into the probability distribution in Eqs. (5.9), (5.10) and (5.11). This is called a linear estimator of the Bloch vector in one-qubit state tomography. Let $\boldsymbol{s}^{\mathrm{L}}$ denote the linear estimator of the Bloch vector, and $\boldsymbol{s}^{\mathrm{L}}_N = (s^{\mathrm{L}}_{N,1}, s^{\mathrm{L}}_{N,2}, s^{\mathrm{L}}_{N,3})^T$ denote the estimate calculated from $\boldsymbol{x}^N$. The explicit form of the estimate is given by

$$s^{\mathrm{L}}_{N,1} = 3\{\nu_N(1, +1) - \nu_N(1, -1)\}, \tag{5.12}$$

$$s^{\mathrm{L}}_{N,2} = 3\{\nu_N(2, +1) - \nu_N(2, -1)\}, \tag{5.13}$$

$$s^{\mathrm{L}}_{N,3} = 3\{\nu_N(3, +1) - \nu_N(3, -1)\}. \tag{5.14}$$

The corresponding estimate of the density matrix is given by $\hat{\rho}(\boldsymbol{s}^{\mathrm{L}}_N) = \frac{1}{2}(\hat{\mathbb{1}} + \boldsymbol{s}^{\mathrm{L}}_N \cdot \boldsymbol{\sigma})$.

The linear estimator is easy to understand and calculate, but there is the problem that the estimates can be unphysical. For example, when $N = 3$ and we obtain the data $\boldsymbol{x}^{N=3} = \{(1, +1), (2, +1), (3, +1)\}$, the linear estimate is $\boldsymbol{s}^{\mathrm{L}}_{N=3} = (1, 1, 1)^T$. In this case, $\|\boldsymbol{s}^{\mathrm{L}}_{N=3}\|_2 = \sqrt{3} > 1$ and the estimated vector is not included in the Bloch ball. A simple solution of this problem is to normalize estimates when their norm is larger than 1, i.e.,

$$s^{\mathrm{NM}}_N := \begin{cases} \boldsymbol{s}^{\mathrm{L}}_N & (\|\boldsymbol{s}^{\mathrm{L}}_N\|_2 \le 1) \\ \boldsymbol{s}^{\mathrm{L}}_N / \|\boldsymbol{s}^{\mathrm{L}}_N\|_2 & (\|\boldsymbol{s}^{\mathrm{L}}_N\|_2 > 1) \end{cases}. \tag{5.15}$$

This is called the norm-minimization estimator with respect to ℓ_2-norm because Eq. 5.15 can be rewritten as

$$s_N^{\mathrm{NM}} = \operatorname*{argmin}_{s' \in B} \|s' - s_N^{\mathrm{L}}\|_2. \tag{5.16}$$

Thus the ℓ_2-norm-minimization estimator chooses as the estimate the Bloch vector closest to the linear estimate with respect to the ℓ_2-norm.

An other solution is a maximum-likelihood estimator s^{ML}. The estimate is defined as

$$s_N^{\mathrm{ML}} := \operatorname*{argmax}_{s' \in B} p(x^N | \Pi^{(6)}, \hat{\rho}(s')). \tag{5.17}$$

The estimate is the Bloch vector such that the probability for the obtained data x^N is maximal in the Bloch ball. Equation (5.17) can be written as

$$\begin{aligned} s_N^{\mathrm{ML}} &= \operatorname*{argmax}_{s' \in B} \prod_{n=1}^{N} p(x_n | \Pi^{(6)}, \hat{\rho}(s')) && (5.18) \\ &= \operatorname*{argmin}_{s' \in B} K(p(\cdot|\Pi^{(6)}, \hat{\rho}(s_N^{\mathrm{L}})) \| p(\cdot|\Pi^{(6)}, \hat{\rho}(s')), && (5.19) \end{aligned}$$

where $K(p\|q) := \sum_{x \in \mathscr{X}} p(x) \ln \frac{p(x)}{q(x)}$ is the Kullback-Leibler divergence of probability distributions p and q. It is called the relative entropy in physics. For s_N^{ML}, as the estimate we choose the Bloch vector closest to the linear estimate with respect to the relative entropy.

4. **Loss function**

In quantum state tomography, the following two loss functions are often used for evaluating estimation errors.

$$\Delta^{\mathrm{IF}}(\rho, \rho') := 1 - \mathrm{Tr}\left[\sqrt{\sqrt{\hat{\rho}}\hat{\rho}'\sqrt{\hat{\rho}}}\right]^2, \tag{5.20}$$

$$\Delta^{\mathrm{T}}(\rho, \rho') := \frac{1}{2}\mathrm{Tr}[|\hat{\rho} - \hat{\rho}'|]. \tag{5.21}$$

Δ^{IF} and Δ^{T} are called the infidelity and the trace distance, respectively. The above definitions are for the arbitrary dimensional case. When the system is a qubit, these loss functions are represented in terms of Bloch vectors as [13]

$$\Delta^{\mathrm{IF}}(s, s') = \frac{1}{2}\left(1 - s \cdot s' - \sqrt{1 - \|s\|^2}\sqrt{1 - \|s'\|^2}\right), \tag{5.22}$$

$$\Delta^{\mathrm{T}}(s, s') = \frac{1}{2}\|s - s'\|_2. \tag{5.23}$$

In Eq. (5.23), the trace distance is represented as half of the ℓ_2-norm, but this is only for a qubit system. In systems with $d \geq 3$, the trace distance includes more than third order terms and cannot be represented as a constant times the ℓ_2-norm.

For a chosen loss function, the expected loss or error probability is used as a figure of merit.

5.1.2.2 General Case

In Sect. 5.1.2.1, we explained the estimation setting for one-qubit state tomography. Here we explain the estimation setting of quantum state tomography for more general systems. The contents are almost same, except for two different points: (i) the dimension d is arbitrary, (ii) the POVM $\boldsymbol{\Pi}$ is not necessarily a projective measurement.

1. **Parametrization**
The Bloch parametrization can be generalized to higher dimensional systems. Let us consider a d-dimensional Hilbert space $\mathscr{H}$, $\hat{\rho} \in \mathscr{S}(\mathscr{H})$, and introduce the generators of $SU(d)$. Suppose that $\boldsymbol{\sigma} = (\hat{\sigma}_1, \ldots, \hat{\sigma}_{d^2-1})$ and $\hat{\sigma}_\alpha$ satisfies

   a. (Hermiticity) $\hat{\sigma}_\alpha = \hat{\sigma}_\alpha^\dagger$.
   b. (Tracelessness) $\mathrm{Tr}[\hat{\sigma}_\alpha] = 0$.
   c. (Orthogonality) $\mathrm{Tr}[\hat{\sigma}_\alpha \hat{\sigma}_\beta] = 2\delta_{\alpha\beta}$.

   By using these matrices, a density matrix $\hat{\rho}$ can be parametrized as [14, 15]

$$\hat{\rho}(\boldsymbol{s}) = \frac{1}{d}\hat{\mathbb{1}} + \frac{1}{2}\boldsymbol{s} \cdot \boldsymbol{\sigma}. \tag{5.24}$$

   The vector

$$s_\alpha = \mathrm{Tr}[\hat{\rho}\hat{\sigma}_\alpha],\ \alpha = 1, \ldots, d^2 - 1, \tag{5.25}$$

   is called the generalized Bloch vector and the R.H.S. of Eq. (5.24) is called the generalized Bloch representation or the generalized Bloch parametrization. A generalizad Bloch vector $\boldsymbol{s}$ is a vector in $(d^2 - 1)$-dimensional Euclidean space and is restricted to a sphere with radius $\sqrt{\frac{2(d-1)}{d}}$, i.e., $\|\boldsymbol{s}\| \leq \sqrt{\frac{2(d-1)}{d}}$. In addition, there are $(d - 2)$ constraints on $\boldsymbol{s}$ coming from positives semidefiniteness, and the total number of constraints is $d - 1$. Let B_d denote the physical region in the parameter space. When $d = 2$, the number of constraint is $d - 1 = 1$ and the constraint can be simply written as $\|\boldsymbol{s}\|_2 \leq 1$, but in higher dimensions the parameter region corresponding to density matrices is only part of the hypersphere. This is a major difference from 2-dimensional systems.
2. **Experimental design**
Commonly, in all types of quantum tomography, the whole of the known components in a tomographic experiment is called the tester. A tester is called "complete"

if the tester makes it possible to completely identify the estimation object in the tomography from the data set of an infinite number of measurement trials, and is called "incomplete" otherwise. The completeness of a tester corresponds to the identifiability by experimental design explained in Sect. 3.1.2. The conditions required for complete testers depends on the type of the estimation object.

In quantum state tomography, we assume that a sequence of copies of an unknown quantum state are given. We perform known quantum measurements on these copies. Hence testers in quantum state tomography are quantum measurements. In standard setting of quantum tomography we choose a combination of measurements. Let $\breve{\boldsymbol{\Pi}} = \{\boldsymbol{\Pi}^{(j)}\}_{j=1}^{J}$ denote a finite set of POVMs. Suppose that for estimating $\hat{\rho}$ we independently perform a measurement described by a POVM $\boldsymbol{\Pi}^{(j)} = \{\hat{\Pi}_m^{(j)}\}_{m\in\mathscr{X}^{(j)}}$ a number $n^{(j)}$ of times ($j = 1, \ldots, J$, $\mathscr{X}^{(j)} = \{1, \ldots, M^{(j)}\}$). The total number of measurement trials is $\sum_{j=1}^{J} n^{(j)} = N$. We define $r^{(j)} := N/n^{(j)}$. Let us define a random variable $X^{(j)}$ as a random variable with a probability distribution as same as $\boldsymbol{\Pi}^{(j)}$. Therefore, deciding on an experimental design X^{exp} is equivalent to deciding on a set of POVMs $\breve{\boldsymbol{\Pi}}$. The necessary and sufficient conditions for a POVM to be a complete tester is that the POVM be a basis of $\mathscr{S}(\mathscr{H})$, and if this condition is satisfied, the tester is called informationally complete. In the same way as density matrices, elements of the POVM can also be parametrized as

$$\hat{\Pi}_m^{(j)} = a_{m,0}^{(j)} \hat{\mathbb{1}} + \boldsymbol{a}_m^{(j)} \cdot \boldsymbol{\sigma}, \tag{5.26}$$

where the real parameters $a_{m,0}^{(j)} = \mathrm{Tr}[\hat{\Pi}_m^{(j)}]/d$ and $a_{m,\beta}^{(j)} = \mathrm{Tr}[\hat{\Pi}_m^{(j)} \sigma_\beta]/2$ should satisfy the normalization conditions ($\sum_{m\in\mathscr{X}} \hat{\Pi}_m = \hat{\mathbb{1}}$), i.e.,

$$\sum_m a_{m,0} = 1, \ \sum_m \boldsymbol{a}_m = \boldsymbol{0}, \tag{5.27}$$

as well as constraints required for positive semidefiniteness. When we perform the measurement on a system in a state described by $\hat{\rho}$, the probability that we observe an outcome m is given by

$$p(m|\boldsymbol{\Pi}^{(j)}, \hat{\rho}) = \mathrm{Tr}\left[\hat{\Pi}_m^{(j)} \hat{\rho}\right] = a_{m,0}^{(j)} + \boldsymbol{a}_m^{(j)} \cdot \boldsymbol{s}. \tag{5.28}$$

Let us define $\boldsymbol{a}_0$ as a vector with (j, m)-th element $a_{m,0}^{(j)}$ and Λ as a matrix with $[(j, m), \alpha]$-th element $a_{m,\alpha}^{(j)}$ ($\alpha = 1, \ldots, d^2 - 1$). We also define $\boldsymbol{p}(\hat{\rho})$ as a vector with (j, m)-th element $p(m|\boldsymbol{\Pi}^{(j)}, \hat{\rho})$. Then Eq. (5.28) can be rewritten as

$$\boldsymbol{p}(\hat{\rho}) = \boldsymbol{a}_0 + \Lambda \boldsymbol{s}. \tag{5.29}$$

The informational completeness of $\breve{\boldsymbol{\Pi}}$ is equivalent to Λ being full rank.

3. **Estimator**

In quantum tomography, the word "reconstruction" is often used with the same meaning of estimation, and "reconstruction scheme" is used with "estimator". In this thesis, we use the words estimation and estimator for consistency with statistical parameter estimation terminology. As explained in Sect. 3.1.2, an estimator is a map from outcomes obtained by performing an experimental design to estimation objects. There are a lot of proposals for estimators in quantum tomography, but here we focus on three of the most popular estimators.

For simplicity, let us consider quantum state tomography on a Hilbert space $\mathscr{H}$. Suppose that N identical copies of a quantum state, represented as $\hat{\rho}^{\otimes N}$, are given. We do not know the details of $\hat{\rho}$ completely, and our purpose is to estimate it. Suppose that we perform quantum measurements described by a POVM $\boldsymbol{\Pi}^{(j)}$ on each copy a number $n^{(j)}$ of times ($j = 1, \ldots, J$), and we obtain a sequence of measurement outcomes $\boldsymbol{x}^N = \{x_1^{(j)}, \ldots, x_{n^{(j)}}^{(j)}\}_{j=1}^J$. Let $\nu_m^{(j)}$ denote the relative frequency of an outcome m in $n^{(j)}$ outcomes of $\boldsymbol{\Pi}^{(j)}$. We define $\boldsymbol{\nu}^{(j)}$ and $\boldsymbol{\nu}_N$ as a vector with m-th element $\nu_m^{(j)}$ and a vector with (j, m)-th element $\nu^{(j)}(m)$, respectively. These vectors are related as $\boldsymbol{\nu}_N = (\boldsymbol{\nu}^{(1)}, \ldots, \boldsymbol{\nu}^{(J)})^T$.

- Linear estimator, $\hat{\rho}^{\mathrm{L}}$ [1, 4, 5, 9, 11, 16]
  Linear estimators give us an estimate reproducing the frequencies of the experimental data, that is, the probability distribution calculated from the estimate coincides with the relative frequencies $\boldsymbol{\nu}_N$. The linear estimate $\hat{\rho}_N^{\mathrm{L}}$ is defined as the matrix satisfying

$$\boldsymbol{\nu}_N = \boldsymbol{p}(\hat{\rho}_N^{\mathrm{L}}). \tag{5.30}$$

  The estimate $\hat{\rho}_N^{\mathrm{L}}$ is calculated from Eq. (5.30) by linear inversion. Equation (5.30) does not always have a solution, and even when it does, it is not guaranteed that $\hat{\rho}_N^{\mathrm{L}}$ is positive semidefinite (Hermiticity and normalization are guaranteed). Therefore a linear estimator can give us unphysical estimation results. In the Bloch parametrization, Eq. (5.30) can be rewritten as

$$\boldsymbol{\nu}_N = \boldsymbol{a}_0 + \Lambda \boldsymbol{s}_N^{\mathrm{L}}. \tag{5.31}$$

  When $\breve{\boldsymbol{\Pi}}$ is informationally complete and Eq. (5.31) has a solution, it is unique and is given by

$$\boldsymbol{s}_N^{\mathrm{L}} = \Lambda_{\mathrm{left}}^{-1}(\boldsymbol{\nu}_N - \boldsymbol{a}_0), \tag{5.32}$$

  where $\Lambda_{\mathrm{left}}^{-1} := (\Lambda^T \Lambda)^{-1} \Lambda^T$ is the left-inverse matrix of Λ.
- Norm-minimization estimator, $\hat{\rho}^{\mathrm{NM}}$
  Let $|||\cdot|||$ denote a matrix norm on $\mathscr{S}(\mathscr{H})$. For a given linear estimate $\hat{\rho}_N^{\mathrm{L}}$, the norm-minimization estimate $\hat{\rho}_N^{\mathrm{NM}}$ is defined as

$$\hat{\rho}_N^{\mathrm{NM}} := \underset{\hat{\rho}' \in \mathscr{S}(\mathscr{H})}{\mathrm{argmin}} \; |||\hat{\rho}' - \hat{\rho}_N^{\mathrm{L}}|||. \tag{5.33}$$

When $\hat{\rho}_N^{\mathrm{L}}$ is in $\mathscr{S}(\mathscr{H})$, $\hat{\rho}_N^{\mathrm{NM}} = \hat{\rho}_N^{\mathrm{L}}$ holds.

- Maximum-likelihood estimator, $\hat{\rho}^{\mathrm{ML}}$ [2, 7, 10]
  A maximum-likelihood estimate is defined as

$$\hat{\rho}_N^{\mathrm{ML}} := \underset{\hat{\rho}' \in \mathscr{S}(\mathscr{H})}{\operatorname{argmax}} \; p(\boldsymbol{x}^N | \breve{\boldsymbol{\Pi}}, \hat{\rho}'). \tag{5.34}$$

  When $n^{(j)} = n$, Eq. (5.34) is rewritten as

$$\hat{\rho}_N^{\mathrm{ML}} = \underset{\hat{\rho}' \in \mathscr{S}(\mathscr{H})}{\operatorname{argmin}} \; K(\boldsymbol{\nu}_N \| \boldsymbol{p}(\rho')). \tag{5.35}$$

  Therefore, a maximum-likelihood estimator can be interpreted as a Kullback-Leibler divergence-minimization estimator. When $\hat{\rho}_N^{\mathrm{L}}$ is in $\mathscr{S}(\mathscr{H})$, $\hat{\rho}_N^{\mathrm{ML}} = \hat{\rho}_N^{\mathrm{L}}$ holds.

There are also other estimators, for example, Bayesian estimator [17, 18] and Maximum-entropy estimator [19].

4. **Figure of merit**
Before discussing figures of merit in quantum tomography, we note the notation of expected loss and error probability. In Chap. 3, we introduced two functions for evaluating estimation errors; expected loss and error probability. They are represented by

$$\bar{\Delta}_N(X^{\mathrm{exp}}, \theta^{\mathrm{est}} | \theta_0) \text{ and } \mathrm{P}^{\Delta}_{\delta, N}(X^{\mathrm{exp}}, \theta^{\mathrm{est}} | \theta_0). \tag{5.36}$$

This is notation for general parameter estimation. As explained above, in quantum state tomography, the estimation object is a density matrix $\hat{\rho}$ and the experimental design X^{exp} is characterized by a set of POVMs $\breve{\boldsymbol{\Pi}}$. We use the following notation for quantum state tomography:

$$\bar{\Delta}_N(\breve{\boldsymbol{\Pi}}, \hat{\rho}^{\mathrm{est}} | \hat{\rho}) \text{ and } \mathrm{P}^{\Delta}_{\delta, N}(\breve{\boldsymbol{\Pi}}, \hat{\rho}^{\mathrm{est}} | \hat{\rho}), \tag{5.37}$$

and

$$\bar{\Delta}_N(A, \boldsymbol{s}^{\mathrm{est}} | \boldsymbol{s}) \text{ and } \mathrm{P}^{\Delta}_{\delta, N}(A, \boldsymbol{s}^{\mathrm{est}} | \boldsymbol{s}) \tag{5.38}$$

for the generalized Bloch representation $\breve{\boldsymbol{\Pi}} = \breve{\boldsymbol{\Pi}}(A)$ and $\hat{\rho} = \hat{\rho}(\boldsymbol{s})$, where $A = (\boldsymbol{a}_0, \Lambda)$.

Let Δ denote a loss function. As explained in the previous subsection, the trace distance and infidelity are often used as loss functions in quantum state tomography, but here we consider a general loss function Δ. In many state tomography experiments, the final goal is to verify a successful preparation of a specific state, say $\hat{\rho}_*$. Suppose that $\hat{\rho}$ is the true density matrix describing the state actually prepared, and $\hat{\rho}_N^{\mathrm{est}}$ denotes an estimate obtained from data. The true state $\hat{\rho}$ is unknown, but we know $\hat{\rho}_*$ because it is the target of the preparation and we also know $\hat{\rho}_N^{\mathrm{est}}$ because it was calculated from experimental data. The quantity that

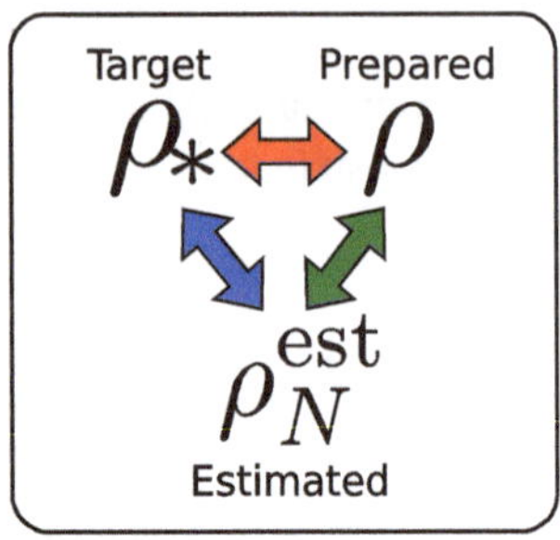

Fig. 5.1 Three-fold relation between target, prepared, and estimated states: $\hat{\rho}_*$ is a target state that an experimentalist tries to prepare, $\hat{\rho}$ is the true prepared state. $\hat{\rho}_N^{\text{est}}$ is an estimate made from N tomographic data sets. (Reproduced from Ref. [55] with permission)

we should evaluate is $\Delta(\hat{\rho}, \hat{\rho}_*)$, i.e., the difference between what we prepared and what we want to prepare. However we do no know $\hat{\rho}$ and so cannot calculate $\Delta(\hat{\rho}, \hat{\rho}_*)$. In many experiments, the accuracy of the preparation is evaluated by the quantity of $\Delta(\hat{\rho}_N^{\text{est}}, \hat{\rho}_*)$ instead of $\Delta(\hat{\rho}, \hat{\rho}_*)$ (see Fig. 5.1). This is not reasonable in general, but when N is sufficiently large, this can be justified as follows. Suppose that the loss function Δ satisfies the triangle inequality. Then we have

$$\Delta(\hat{\rho}, \hat{\rho}_*) \leq \Delta(\hat{\rho}, \hat{\rho}_N^{\text{est}}) + \Delta(\hat{\rho}_N^{\text{est}}, \hat{\rho}_*). \tag{5.39}$$

When N is sufficiently large, the value of $\Delta(\hat{\rho}, \hat{\rho}_N^{\text{est}})$ is negligible. Therefore we have

$$\Delta(\hat{\rho}, \hat{\rho}_*) \lesssim \Delta(\hat{\rho}_N^{\text{est}}, \hat{\rho}_*), \tag{5.40}$$

and evaluating $\Delta(\hat{\rho}_N^{\text{est}}, \hat{\rho}_*)$ is an upper bound for $\Delta(\hat{\rho}, \hat{\rho}_*)$. When the estimation error $\Delta(\hat{\rho}, \hat{\rho}_N^{\text{est}})$ is sufficiently small, the above logic is correct, but in some experiments, N is not sufficiently large and $\Delta(\hat{\rho}, \hat{\rho}_N^{\text{est}})$ is not negligible. In order to evaluate the performance of state preparation in such experiments, we need a different method applicable to the finite sample case.

In this thesis we propose such a method, strictly evaluating $\Delta(\hat{\rho}, \hat{\rho}_*)$. We derive upper bounds Δ_u and P_u on expected loss and error probability, i.e.,

$$\bar{\Delta}_N(\breve{\boldsymbol{\Pi}}, \hat{\rho}^{\text{est}}|\hat{\rho}) \leq \Delta_u(N, \breve{\boldsymbol{\Pi}}, \hat{\rho}^{\text{est}}), \tag{5.41}$$

$$\mathrm{P}^{\Delta}_{\delta,N}(\breve{\boldsymbol{\Pi}}, \hat{\rho}^{\text{est}}|\hat{\rho}) \leq \mathrm{P}^{\Delta}_u(\delta, N, \breve{\boldsymbol{\Pi}}, \hat{\rho}^{\text{est}}), \tag{5.42}$$

for any $\hat{\rho} \in \mathcal{S}(\mathcal{H})$. Using these functions as well as the triangle inequality, we can insist that for any true state $\hat{\rho}$,

$$\Delta(\hat{\rho}, \hat{\rho}_*) \leq \Delta_u(N, \breve{\boldsymbol{\Pi}}, \hat{\rho}^{\text{est}}) + \Delta(\hat{\rho}_N^{\text{est}}, \hat{\rho}_*) \tag{5.43}$$

on average, and

$$\Delta(\hat{\rho}, \hat{\rho}_*) \leq \delta + \Delta(\hat{\rho}_N^{\mathrm{est}}, \hat{\rho}_*) \tag{5.44}$$

with probability greater than $1 - \mathrm{P}_u^{\Delta}(\delta, N, \breve{\boldsymbol{\Pi}}, \hat{\rho}^{\mathrm{est}})$. The former approach using Δ_u might be enough for tomographic experiments which do not require rigorous evaluation of estimation errors. When the purpose of the state preparation is a secure quantum cryptographic protocol, the former is not suitable and the latter approach using P_u^{Δ} should be used.

There are two methods related to finite sample analysis of estimation errors in quantum tomography. One is compressed sensing, and the other is that of confidence regions. In [20], an estimation setting different from that of standard state tomography was considered. When the rank of the density matrix describing the true state is sufficiently small and we know this information, intuitively it would be natural for us to expect that we can reduce our experimental effort for estimation. This intuition proves to be correct. In [20], estimation schemes using informationally incomplete POVMs and two estimators called compressed sensing were proposed, and it was shown that these estimation schemes allowed one to estimate low-rank states with sufficient precision, in the sense that upper-bounds on the error probabilities for the estimation schemes were derived. However, these formulae include indeterminate coefficients, and their value cannot be calculated explicitly. Therefore the bounds are not directly applicable to the evaluation of tomography experiments even if the true state is low-rank. Additionally, the rank of the true state can be large in general, and the results obtained in [20] are not applicable in this case.

In [21, 22], a different approach, that of confidence regions, which is a concept originating from statistics, was investigated. A region estimator, $\Xi = \{\Xi_1, \Xi_2, \ldots\}$, is defined as a set of maps from the data to a subset of $\mathscr{S}(\mathscr{H})$. Let ε denote a positive number satisfying $0 < \varepsilon < 1$. An estimated subset $\Xi_N(\boldsymbol{x}^N)$ is called a region estimate. When a region estimator satisfies

$$\mathbb{P}\Big[\, \hat{\rho} \in \Xi_N(X^N) |\, \hat{\rho} \Big] := \sum_{\boldsymbol{x}^N \in \mathscr{X}^N;\ \hat{\rho} \in \Xi_N(\boldsymbol{x}^N)} p(\boldsymbol{x}^N | \breve{\boldsymbol{\Pi}}, \hat{\rho}) \tag{5.45}$$

$$\geq 1 - \varepsilon \tag{5.46}$$

for any $\hat{\rho} \in \mathscr{S}(\mathscr{H})$, it is called a confidence region estimator with confidence level $1-\varepsilon$ at the number of measurement trials N. A confidence region estimator with $1-\varepsilon$ at N guarantees that for given data $\boldsymbol{x}^N$ the true density matrix $\hat{\rho}$ is included in the region estimate $\Xi_N(\boldsymbol{x}^N)$ with probability at least $1 - \varepsilon$. For a given confidence level $1 - \varepsilon$, constructing a confidence region estimator at N is the statistical estimation problem for confidence regions. For example, $\Xi_N(\boldsymbol{x}^N) = \mathscr{S}(\mathscr{H})$ gives a confidence region estimator with confidence level 1 at any N, but this is trivial. In the confidence region approach, a region estimator which gives smaller confidence regions is better. In [21, 22], confidence region estimators for quantum state estimation were proposed, and their confidence level and volume of the confidence region were analyzed.

For a point estimate, let us define a δ-ball $B_{\Delta\leq\delta}(\hat{\rho}_N^{\mathrm{est}})$ by

$$B_{\Delta\leq\delta}(\hat{\rho}_N^{\text{est}}) := \left\{\hat{\rho}' \in \mathscr{S}(\mathscr{H}) |\ \Delta(\hat{\rho}_N^{\text{est}}, \hat{\rho}') \leq \delta\right\}. \tag{5.47}$$

The error probability can be rewritten as

$$\mathrm{P}^{\Delta}_{\delta,N}(\breve{\boldsymbol{\Pi}}, \hat{\rho}^{\text{est}}|\hat{\rho}) = \mathbb{P}\left[\ \hat{\rho} \notin B_{\Delta\leq\delta}(\hat{\rho}_N^{\text{est}})|\ \hat{\rho}\right]. \tag{5.48}$$

If Eq. (5.42) holds, $B_{\Delta\leq\delta}(\hat{\rho}_N^{\text{est}})$ is a confidence region estimate with confidence $1-\mathrm{P}^{\Delta}_u$ at N. Therefore, a point estimator is easily used for making a region estimator, and in our approach P^{Δ}_u gives the value of the confidence level. On the other hand, if estimates of a confidence region estimator are not calculated by point estimates, as with the region estimators proposed in [21, 22], the confidence region is not directly applicable to the evaluation of quantum tomography experiments, since these conventionally make use of only point estimates.

In Sects. 5.2 and 5.3, we consider the case in which the POVM used is informationally complete and there are no assumptions on the true state, and derive the explicit form of Δ_u and P^{Δ}_u, respectively. These results are directly applicable to the evaluation of current tomography experiments.

5.2 Expected Loss

In the Sect. 5.1, we explained estimation settings in quantum tomography. In this section, we explain our results on expected losses for extended linear, extended norm-minimization, and maximum-likelihood estimators. The goal of this section is to derive a function Δ_u, upper-bounding expected losses for these estimators. In Sect. 5.2.1, we explain two drawbacks of linear estimators, and introduce an extended linear estimator in order to avoid one of these problems. Using the Hoeffding inequality, we derive functions which upper-bound the expected losses for an extended linear estimator. An extended linear estimator can still give unphysical estimates. In Sect. 5.2.2, we propose a new estimator called an extended norm-minimization, which always gives physical estimates. We derive functions upper-bounding the expected losses of those estimators by combining the results of an extended linear estimator explained in Sect. 5.2.1 and some inequalities on norms and loss functions. In Sect. 5.2.3, we analyze the expected losses of a maximum-likelihood estimator by introducing two approximations. We derive approximate expected losses in one-qubit state tomography and evaluate the accuracy of the approximation numerically.

5.2.1 Extended Linear Estimator

In this subsection, we analyze the expected losses of an extended linear estimator in quantum state tomography.

5.2.1.1 Definition

Before the analysis, we clarify the condition that a linear estimate exists. A linear estimate is a solution of the equation

$$\boldsymbol{\nu}_N - \boldsymbol{a}_0 = \Lambda \boldsymbol{s}' \tag{5.49}$$

This is an inhomogeneous equation, whose explanations are in Sect. A.1.2. As explained in Sect. A.1.2, $\boldsymbol{\nu}_N - \boldsymbol{a}_0 \in \mathrm{Im}\,(\Lambda)$ is the necessary and sufficient condition for the existence of the solution of Eq. (5.49). Suppose that $\boldsymbol{\nu}_N - \boldsymbol{a}_0 \in \mathrm{Im}(\Lambda)$ is satisfied. Then the solution of Eq. (5.49) exists. When $\boldsymbol{\Pi}(A)$ is informationally complete, Λ is full-rank and the left-inverse matrix $\Lambda_{\mathrm{left}}^{-1}$ exists. The left-inverse matrix is uniquely determined by $\Lambda_{\mathrm{left}}^{-1} = (\Lambda^T \Lambda)^{-1} \Lambda^T$. Then the solution of Eq. (5.49) is uniquely determined by

$$\boldsymbol{s}' = \Lambda_{\mathrm{left}}^{-1}(\boldsymbol{\nu}_N - \boldsymbol{a}_0). \tag{5.50}$$

When $\breve{\boldsymbol{\Pi}}(A)$ is not informationally complete (IC), the left-inverse matrix does not exists and the solution is not unique. Summarizing the above discussion, we obtain

$$\boldsymbol{s}_N^{\mathrm{L}} = \begin{cases} \text{No solutions} & \text{if } \boldsymbol{\nu}_N - \boldsymbol{a}_0 \notin \mathrm{Im}\,(\Lambda) \\ \Lambda_{\mathrm{left}}^{-1}(\boldsymbol{\nu}_N - \boldsymbol{a}_0) & \text{if } \boldsymbol{\nu}_N - \boldsymbol{a}_0 \in \mathrm{Im}\,(\Lambda),\ \breve{\boldsymbol{\Pi}} \text{is IC} \\ \text{Not unique} & \text{if } \boldsymbol{\nu}_N - \boldsymbol{a}_0 \in \mathrm{Im}\,(\Lambda),\ \breve{\boldsymbol{\Pi}} \text{is not IC} \end{cases} . \tag{5.51}$$

In quantum state tomography, $\breve{\boldsymbol{\Pi}}$ is chosen as informationally complete, and we do not need to consider the third line in Eq. (5.51). Let $\mathscr{P}(\mathscr{X}^{(j)})$ denote the set of probability distributions with the sample space $\mathscr{X}^{(j)}$ $(j = 1, \ldots, J)$. Then, $\boldsymbol{\nu}^{(j)}$ is in $\mathscr{P}(\mathscr{X}^{(j)})$, and $\boldsymbol{\nu}_N = (\boldsymbol{\nu}^{(1)}, \ldots, \boldsymbol{\nu}^{(J)})^T$ is in $\mathscr{R} := \oplus_{j=1}^{J} \mathscr{P}(\mathscr{X}^{(j)})$. We introduce the following notation for sets of probability distributions:

- $\mathscr{R}_{\mathrm{NS}}$: the set of $\boldsymbol{\nu}_N$ such that Eq. (5.49) does not have any solution.
- $\mathscr{R}_{\mathrm{S}}$: the set of $\boldsymbol{\nu}_N$ such that Eq. (5.49) has a solution.
- $\mathscr{R}_{\mathrm{UP}}$: the set of $\boldsymbol{\nu}_N$ such that Eq. (5.49) has a solution and the solution is unphysical.
- $\mathscr{R}_{\mathrm{P}}$: the set of $\boldsymbol{\nu}_N$ such that Eq. (5.49) has a solution and the solution is physical.

By definition, we have

$$\begin{aligned} \mathscr{R} &= \mathscr{R}_{\mathrm{NS}} \oplus \mathscr{R}_{\mathrm{S}} && (5.52) \\ &= \mathscr{R}_{\mathrm{NS}} \oplus \mathscr{R}_{\mathrm{UP}} \oplus \mathscr{R}_{\mathrm{P}}. && (5.53) \end{aligned}$$

Let Δ denote a loss function. When $\boldsymbol{\nu}_N(\boldsymbol{x}^N) \notin \mathscr{R}_{\mathrm{S}}$, the linear estimate does not exist. We cannot calculate the value of the loss function for the outcome sequence $\boldsymbol{x}^N$, and the expected loss is indefinite. To avoid this problem, we introduce a new estimator called an extended linear estimator, $\boldsymbol{s}^{\mathrm{eL}}$, defined by

$$\boldsymbol{s}_N^{\mathrm{eL}}(\boldsymbol{x}^N) := \Lambda_{\mathrm{left}}^{-1}(\boldsymbol{\nu}_N - \boldsymbol{a}_0), \ \forall \boldsymbol{\nu}_N \in \mathscr{R}. \tag{5.54}$$

We define the extended linear estimator of $\hat{\rho}$ by

$$\hat{\rho}_N^{\mathrm{eL}} := \hat{\rho}(\boldsymbol{s}_N^{\mathrm{eL}}). \tag{5.55}$$

From a straightforward calculation, we can prove that $\hat{\rho}_N^{\mathrm{eL}}$ is a linear least squares estimator, i.e., it satisfies

$$\hat{\rho}_N^{\mathrm{eL}} = \operatorname*{argmin}_{\substack{\sigma;\sigma=\sigma^\dagger, \\ \mathrm{Tr}[\sigma]=1}} \|\boldsymbol{p}(\sigma) - \boldsymbol{\nu}_N\|_2. \tag{5.56}$$

We explain some work related to finite sample analysis of expected losses in quantum state tomography. In [23], an estimator that is a linear function of relative frequencies and always has a solution was introduced. For this estimator and the Hilbert-Schmidt distance, lower-bounds of the pointwise and average expected losses were derived. In [24], the lower-bounds and the approximate behavior of these expected losses were analyzed for a specific design of experiment. The results obtained in [23, 24] are lower-bounds on expected losses. Therefore the results are not applicable to the evaluation of estimation errors in quantum tomography experiments. In the next subsection, we focus on the extended linear estimator. In contrast to the previous works, we derive functions upper-bounding the maximum value of $\bar{\Delta}_N(A, \boldsymbol{s}^{\mathrm{eL}}|\boldsymbol{s})$ for general informationally complete sets of POVMs.

5.2.1.2 Analysis of Expected Losses

For three vector distances ℓ_1, ℓ_2, and ℓ_∞, we have the following theorem.

Theorem 5.1 (Expected loss, $\boldsymbol{s}^{\mathrm{eL}}$, ℓ_1–, ℓ_2–, ℓ_∞ -distances)
When we choose the square of the ℓ_1-, ℓ_2-, and ℓ_∞-distances as the loss function, we have

$$\bar{\Delta}_N^{(\ell_1)^2}(A, \boldsymbol{s}^{\mathrm{eL}}|\boldsymbol{s}) \le \frac{d^2-1}{N} \max_{\boldsymbol{p}\in\mathscr{P}_{\mathrm{P}}(\mathscr{X})} \mathrm{tr}[(\Lambda_{\mathrm{left}}^{-1})^T \Lambda_{\mathrm{left}}^{-1} \Gamma(\boldsymbol{p})], \tag{5.57}$$

$$\bar{\Delta}_N^{(\ell_2)^2}(A, \boldsymbol{s}^{\mathrm{eL}}|\boldsymbol{s}) \le \frac{1}{N} \max_{\boldsymbol{p}\in\mathscr{P}_{\mathrm{P}}(\mathscr{X})} \mathrm{tr}[(\Lambda_{\mathrm{left}}^{-1})^T \Lambda_{\mathrm{left}}^{-1} \Gamma(\boldsymbol{p})], \tag{5.58}$$

$$\bar{\Delta}_N^{(\ell_\infty)^2}(A, \boldsymbol{s}^{\mathrm{eL}}|\boldsymbol{s}) \le \frac{1}{N} \max_{\boldsymbol{p}\in\mathscr{P}_{\mathrm{P}}(\mathscr{X})} \mathrm{tr}[(\Lambda_{\mathrm{left}}^{-1})^T \Lambda_{\mathrm{left}}^{-1} \Gamma(\boldsymbol{p})]. \tag{5.59}$$

Proof (Theorem 5.1)
First, we prove Eq. (5.58). From Eq. (3.36), we obtain

$$\bar{\Delta}_N^{(\ell_2)^2}(A, s^{\mathrm{eL}}|s) = \mathbb{E}_s[\|s_N^{\mathrm{eL}} - s\|_2^2] \tag{5.60}$$
$$= \mathrm{tr}[(\Lambda_{\mathrm{left}}^{-1})^T \Lambda_{\mathrm{left}}^{-1} \mathbb{E}_s[(\nu_N - p)(\nu_N - p)^T]] \tag{5.61}$$
$$= \frac{1}{N}\mathrm{tr}[(\Lambda_{\mathrm{left}}^{-1})^T \Lambda_{\mathrm{left}}^{-1} \Gamma(p)] \tag{5.62}$$
$$\leq \frac{1}{N} \max_{p \in \mathscr{P}_{\mathrm{P}}(\mathscr{X})} \mathrm{tr}[(\Lambda_{\mathrm{left}}^{-1})^T \Lambda_{\mathrm{left}}^{-1} \Gamma(p)]. \tag{5.63}$$

Equations (5.57) and (5.59) are obtained by combining Eq. (5.58) with Eqs. (5.162) and (5.164). □

Next, let us consider the square of the Hilbert-Schmidt distance and square of the trace distance. We have the following theorem.

Theorem 5.2 (Expected loss, $\hat{\rho}^{\mathrm{eL}}$, Δ^{HS2}, Δ^{T2})
When we choose the square of the Hilbert-Schmidt distance and the square of the trace distance, we have

$$\bar{\Delta}_N^{\mathrm{HS2}}(\breve{\boldsymbol{\Pi}}, \hat{\rho}^{\mathrm{eL}}|\hat{\rho}) \leq \frac{1}{4N} \max_{p \in \mathscr{P}_{\mathrm{P}}(\mathscr{X})} \mathrm{tr}[(\Lambda_{\mathrm{left}}^{-1})^T \Lambda_{\mathrm{left}}^{-1} \Gamma(p)], \tag{5.64}$$
$$\bar{\Delta}_N^{\mathrm{T2}}(\breve{\boldsymbol{\Pi}}, \hat{\rho}^{\mathrm{eL}}|\hat{\rho}) \leq \frac{d}{8N} \max_{p \in \mathscr{P}_{\mathrm{P}}(\mathscr{X})} \mathrm{tr}[(\Lambda_{\mathrm{left}}^{-1})^T \Lambda_{\mathrm{left}}^{-1} \Gamma(p)]. \tag{5.65}$$

Proof (Theorem 5.2)
Equation (5.64) can be obtained from Eq. (5.58) and the equality

$$\Delta^{\mathrm{HS}}(\hat{\rho}(s), \hat{\rho}(s')) = \frac{1}{2}\|s - s'\|_2. \tag{5.66}$$

From Eq. (5.166), we obtain

$$\Delta^{\mathrm{T}}(\hat{\rho}, \hat{\rho}') \leq \sqrt{\frac{d}{2}} \Delta^{\mathrm{HS}}(\hat{\rho}, \hat{\rho}'). \tag{5.67}$$

By combining Eq. (5.67) and (5.58) , we obtain Eq. (5.65). □

An extended linear estimate can be unphysical, and in the cases we cannot calculate the infidelity $\Delta^{\mathrm{IF}}(\hat{\rho}^{\mathrm{eL}}, \hat{\rho})$ because infidelity is defined only for positive semidefinite matrices (the trace and the Hilbert-Schmidt distances are applicable for any Hermitian matrices that are not necessary positive semidefinite). Therefore the expected infidelity for an extended linear estimator does not exist.

5.2.2 Extended Norm-Minimization Estimator

In this subsection, we analyze the expected losses for the extended ℓ_2-norm-minimization estimator in quantum state tomography. As in the previous subsection, we define the extended norm-minimization estimator $s^{\ell_2\text{-eNM}}$ by

$$s^{\ell_2\text{-eNM}}(x^N) := \underset{s' \in B_d}{\operatorname{argmin}} \|s' - s_N^{\mathrm{eL}}(x^N)\|_2, \ \forall \boldsymbol{\nu}_N \in \mathscr{R}. \tag{5.68}$$

First, let us consider the square of the ℓ_2-distance as the loss function. Then the following theorem holds.

Theorem 5.3 (Expected loss, $s^{\ell_2\text{-eNM}}, \ell_2$)

$$\bar{\Delta}_N(A, s^{\ell_2\text{-eNM}}|s) \le \frac{1}{N} \max_{p \in \mathscr{P}_{\mathrm{P}}(\mathscr{X})} \mathrm{tr}[(\Lambda_{\mathrm{left}}^{-1})^T \Lambda_{\mathrm{left}}^{-1} \Gamma(p)]. \tag{5.69}$$

Proof (Theorem 5.3)
By applying Theorem 5.13 to $s^{\ell_2\text{-eNM}}$, we obtain

$$\|s_N^{\ell_2\text{-eNM}} - s\|_2 \le \|s_N^{\mathrm{eL}} - s\|_2. \tag{5.70}$$

Therefore we obtain

$$\bar{\Delta}_N^{(\ell_2)^2}(A, s^{\ell_2\text{-eNM}}|s) \le \bar{\Delta}_N^{(\ell_2)^2}(A, s^{\mathrm{eL}}|s) \tag{5.71}$$

$$\le \frac{1}{N} \max_{p \in \mathscr{P}_{\mathrm{P}}(\mathscr{X})} \mathrm{tr}[(\Lambda_{\mathrm{left}}^{-1})^T \Lambda_{\mathrm{left}}^{-1} \Gamma(p)]. \tag{5.72}$$

□

By combining the same technique used in the proofs of Theorems 5.1 and 5.2 with Theorem 5.3, we can show that the expected losses considered in the theorems for the extended ℓ_2-norm-minimization estimator have the same upper bounds.

Theorem 5.4 (Expected loss, $\hat{\rho}^{\ell_2\text{-eNM}}, \Delta^{(\ell_1)^2}, \Delta^{(\ell_\infty)^2}, \Delta^{\mathrm{HS2}}, \Delta^{\mathrm{T2}}, \Delta^{\mathrm{IF2}}$)
The following inequalities hold:

$$\bar{\Delta}_N^{(\ell_1)^2}(A, s^{\ell_2\text{-eNM}}|s) \le \frac{d^2-1}{N} \max_{p \in \mathscr{P}_{\mathrm{P}}(\mathscr{X})} \mathrm{tr}[(\Lambda_{\mathrm{left}}^{-1})^T \Lambda_{\mathrm{left}}^{-1} \Gamma(p)], \tag{5.73}$$

$$\bar{\Delta}_N^{(\ell_\infty)^2}(A, s^{\ell_2\text{-eNM}}|s) \le \frac{1}{N} \max_{p \in \mathscr{P}_{\mathrm{P}}(\mathscr{X})} \mathrm{tr}[(\Lambda_{\mathrm{left}}^{-1})^T \Lambda_{\mathrm{left}}^{-1} \Gamma(p)], \tag{5.74}$$

$$\bar{\Delta}_N^{\mathrm{HS2}}(\breve{\boldsymbol{\Pi}}, \hat{\rho}^{\ell_2\text{-eNM}}|\hat{\rho}) \le \frac{1}{4N} \max_{p \in \mathscr{P}_{\mathrm{P}}(\mathscr{X})} \mathrm{tr}[(\Lambda_{\mathrm{left}}^{-1})^T \Lambda_{\mathrm{left}}^{-1} \Gamma(p)], \tag{5.75}$$

$$\bar{\Delta}_N^{\mathrm{T2}}(\breve{\boldsymbol{\Pi}}, \hat{\rho}^{\ell_2\text{-eNM}}|\hat{\rho}) \le \frac{d}{8N} \max_{p \in \mathscr{P}_{\mathrm{P}}(\mathscr{X})} \mathrm{tr}[(\Lambda_{\mathrm{left}}^{-1})^T \Lambda_{\mathrm{left}}^{-1} \Gamma(p)]. \tag{5.76}$$

$$\bar{\Delta}_N^{\mathrm{IF2}}(\breve{\boldsymbol{\Pi}}, \hat{\rho}^{\ell_2\text{-eNM}}|\hat{\rho}) \leq \frac{d}{2N} \max_{\boldsymbol{p}\in\mathscr{P}_{\mathrm{P}}(\mathscr{X})} \mathrm{tr}[(\Lambda_{\mathrm{left}}^{-1})^T \Lambda_{\mathrm{left}}^{-1} \Gamma(\boldsymbol{p})]. \tag{5.77}$$

5.2.3 Maximum-Likelihood Estimator

In this subsection, we analyze the behavior of expected losses of a maximum-likelihood estimator with finite data. As explained in Sect. 3.2, a maximum-likelihood estimator attains the Cramér-Rao bound asymptotically, and therefore for sufficiently large N, the mean squared error decreases as $\mathrm{tr}[F(A, \boldsymbol{s})^{-1}]/N + O(1/N^{3/2})$. The Cramér-Rao bound is often used to evaluate the estimation errors of a maximum-likelihood estimator, but there are problems applying the bound to evaluating the expected losses for finite data sets. The inequality holds only for a specific class of estimators, namely those that are unbiased. A maximum-likelihood estimator is asymptotically unbiased, but is not unbiased for finite N, so the expected losses can be smaller than the bound for finite N. Especially when the purity of the true density matrix is high, the bias becomes larger. This is due to the boundary in the parameter space imposed by the condition that density matrices be positive semidefinite, and the expected losses can deviate significantly from the asymptotic behavior [25, 26]. A natural question is then to ask at what value of N the expected losses begin to behave asymptotically. If N is large enough for the effect of the bias to be negligible, we can safely apply the asymptotic theory for evaluating the estimation error in an experiment. However, in general, determining the effects of the bias is a difficult problem because a maximum-likelihood estimator is a non-linear map from data to parameter. As the first step for understanding the effect of the bias, we introduce two approximations and try to understand the effect roughly. For simplicity, we focus on one-qubit state tomography. In Sect. 5.2.3.1, we analyze the bias effect theoretically. Applying ideas from classical statistical estimation theory, we derive an approximate form of the expected losses for finite N. In Sect. 5.2.3.2, we analyze the bias effect numerically, giving the results of our pseudo-random numerical experiments. These indicate that the function we derived reproduces the behavior of the expected losses for finite N more precisely than the Cramér-Rao bound. This makes it possible to predict the point at which the behavior of the expected infidelity becomes effectively asymptotic.

5.2.3.1 Theoretical Analysis

In this section, we derive a function which approximates the expected losses of the squared Hilbert-Schmidt distance and infidelity for finite data sets.

We consider one-qubit state tomography. Let $\boldsymbol{s}$ denote the true Bloch vector and $\boldsymbol{\Pi}$ denote an informationally complete POVM on $\mathbb{C}^2$. It is not necessarily the 6-state POVM. Suppose that for a given sequence of outcomes $\boldsymbol{x}^N$ there exists a linear estimate $\boldsymbol{s}_N^{\mathrm{L}}$. Then the maximum-likelihood estimate is calculated as

$$s_N^{\mathrm{ML}} = \begin{cases} s_N^{\mathrm{L}} & (s_N^{\mathrm{ML}} \in B) \\ \mathrm{argmin}_{s' \in B} \Delta^{\mathrm{KL}}(s_N^{\mathrm{L}}, s') & (s_N^{\mathrm{ML}} \notin B) \end{cases}, \tag{5.78}$$

where Δ^{KL} is the loss function on the parameter space induced by the Kullback-Leibler divergence,

$$\Delta^{\mathrm{KL}}(s, s') := K(p_{A,s} \| p_{A,s'}). \tag{5.79}$$

The expected loss of s_N^{L} was analyzed in Sect. 5.2.1. The difference of s_N^{ML} from s_N^{L} occurs when s^{L} is unphysical. In order to analyze the expected losses of a maximum-likelihood estimator, it is important to analyze the behavior of s_N^{ML} when s_N^{L} is unphysical. The maximum-likelihood estimate is a non-linear function of s_N^{L} because of the minimization in Eq. (5.78), and it is difficult to exactly analyze the behavior. In this subsubsection, we try to derive not the exact form but a simpler function which reproduces the behavior of the true function accurately enough to help us understand the bias effect. In order to accomplish this, we introduce two approximations. First, we approximate the multinomial distribution generated by successive trials by a Gaussian distribution. Second, we approximate the spherical boundary by a plane tangent to its boundary.

1. **Two approximations**
From the central limit theorem, we can readily prove that the distribution of a linear estimator s^{L} converges to a Gaussian distribution with mean s and covariance matrix $F(A, s)^{-1}$. For finite N, we approximate the true probability distribution by the Gaussian distribution

$$p_G(s_N^{\mathrm{L}}|s) := \frac{N^{3/2}}{(2\pi)^{3/2}\sqrt{\det F(A, s)^{-1}}} \exp\left[-\frac{N}{2}(s_N^{\mathrm{L}} - s) \cdot F(A, s)(s_N^{\mathrm{L}} - s)\right]. \tag{5.80}$$

We will refer to this as the Gaussian distribution approximation (GDA). Because the approximation of the multinomial distribution by the GDA becomes better as each outcome probability grows sufficiently larger than 0, the expected losses under the GDA should be closer to the true expected losses the farther the true Bloch vector is from alignment with the axes in the Bloch sphere defined by the measurement.

For a one-qubit system, the boundary between the physical and unphysical regions of the state space is a sphere with unit radius. Despite its simplicity, it is difficult to derive the explicit formula for a maximum-likelihood estimator even in this case. Indeed, this is a major contributor to the general complexity of the expected loss behavior in quantum tomography. We therefore choose the simplest possible way to approximate the boundary, namely by replacing it with a plane in the state space. Suppose that the true Bloch vector is $s \in B$. The boundary of the Bloch ball, ∂B, is represented as

$$\partial B := \{s' \in \mathbb{R}^3 | \, \|s'\|_2 = 1\}. \tag{5.81}$$

We approximate this by the tangent plane to the sphere at the point $\boldsymbol{e}_s := \boldsymbol{s}/\|\boldsymbol{s}\|_2$, represented as

$$\partial D_s := \{\boldsymbol{s}' \in \mathbb{R}^3 |\ \boldsymbol{s} \cdot (\boldsymbol{s}' - \boldsymbol{e}_s) = 0\}, \tag{5.82}$$

and so the approximated parameter space is represented as

$$D_s = \{\boldsymbol{s}' \in \mathbb{R}^3 |\ \boldsymbol{s} \cdot (\boldsymbol{s}' - \boldsymbol{e}_s) \leq 0\}. \tag{5.83}$$

We will refer to this as the linear boundary approximation (LBA). The LBA is a specific case of tangent cone methods in statistical estimation theory which have been developed and used for analyzing models with constrained parameters in classical statistical estimation theory [27, 28]. It is known that the distribution of a maximum-likelihood estimator in a constrained parameter estimation problem converges to the Gaussian distribution with the boundary approximated by a tangent cone, as well as that the dominant term in the approximate expected loss is equivalent to that in the actual one [28]. Therefore it is guaranteed that the expected losses approximated by the GDA and LBA converge to their true values in the limit of infinite data, and the dominant term in the approximate expected loss coincides with that in the actual expected loss.

2. **Approximated maximum-likelihood estimator**
In [28], it is proved that $\sqrt{N}(\boldsymbol{s}_N^{\mathrm{ML}} - \boldsymbol{s})$ converges to $\sqrt{N}(\tilde{\boldsymbol{s}}_N^{\mathrm{ML}} - \boldsymbol{s})$ in distribution as $N \to \infty$, where

$$\tilde{\boldsymbol{s}}_N^{\mathrm{ML}} := \mathrm{argmin}_{\boldsymbol{s}' \in \mathrm{D}_s} (\boldsymbol{s}_{\mathrm{N}}^{\mathrm{L}} - \boldsymbol{s}') \cdot \mathrm{F}(\mathrm{A}, \boldsymbol{s})(\boldsymbol{s}_{\mathrm{N}}^{\mathrm{L}} - \boldsymbol{s}') \tag{5.84}$$

is the maximum-likelihood estimator for the Gaussian distribution p_G with the linearized parameter space D_s. Roughly speaking, this convergence in distribution guarantees that the asymptotic behavior of the dominant term of the approximated estimator $\tilde{\boldsymbol{s}}^{\mathrm{ML}}$ is equivalent to that of the true maximum-likelihood estimator $\boldsymbol{s}^{\mathrm{ML}}$. As explained in the previous paragraph, this asymptotic equivalence justifies the use of the two approximations, GDA and LBA, for the analysis of a maximum-likelihood estimator. By using the Lagrange multiplier method, we can derive the approximate maximum-likelihood estimates as

$$\tilde{\boldsymbol{s}}_N^{\mathrm{ML}} = \begin{cases} \boldsymbol{s}_N^{\mathrm{L}} & (\boldsymbol{s}_N^{\mathrm{L}} \in D_s) \\ \boldsymbol{s}_N^{\mathrm{L}} - \frac{\boldsymbol{e}_s \cdot \boldsymbol{s}_N^{\mathrm{L}} - 1}{\boldsymbol{e}_s \cdot F(A,\boldsymbol{s})^{-1} \boldsymbol{e}_s} F(A, \boldsymbol{s})^{-1} \boldsymbol{e}_s & (\boldsymbol{s}_N^{\mathrm{L}} \notin D_s) \end{cases}. \tag{5.85}$$

We note that $\tilde{\boldsymbol{s}}_N^{\mathrm{ML}}$ depends on the true parameter $\boldsymbol{s}$, and so by definition it is not an estimator—it is a vector introduced for the purpose of approximating expected losses of a maximum-likelihood estimator. Intuitively, it takes the value of the linear estimate if that estimate is physical, and if it is unphysical a correction vector is added to bring it back within the physical region.

3. **Expected squared Hilbert-Schmidt distance**
From a straightforward calculation using formulas for Gaussian integrals, we can derive the approximate expected squared Hilbert-Schmidt distance.

$$\bar{\Delta}_N^{\text{HS2}}(\tilde{s}^{\text{ML}}|s) = \frac{1}{4N}\left(\text{tr}[F(A,s)^{-1}] - \frac{1}{2}\frac{e_s \cdot F(A,s)^{-2} e_s}{e_s \cdot F(A,s)^{-1} e_s}\text{erfc}\left[\sqrt{\frac{N}{N^*}}\right]\right) - \frac{1}{4}\frac{1-\|s\|_2}{\sqrt{2\pi e_s \cdot F(A,s)^{-1} e_s}}\frac{e_s \cdot F(A,s)^{-2} e_s}{e_s \cdot F(A,s)^{-1} e_s}\frac{e^{-N/N^*}}{\sqrt{N}} + \frac{1}{8}\{1-(\|s\|_2)^2\}\frac{e_s \cdot F(A,s)^{-2} e_s}{(e_s \cdot F(A,s)^{-1} e_s)^2}\text{erfc}\left[\sqrt{\frac{N}{N^*}}\right], \tag{5.86}$$

where

$$\text{erfc}[a] := \frac{2}{\sqrt{\pi}}\int_a^\infty dt\, e^{-t^2} \tag{5.87}$$

is the complementary error function and

$$N^* := 2\frac{e_s \cdot F(A,s)^{-1} e_s}{(1-\|s\|_2)^2} \tag{5.88}$$

is a typical scale for the number of trials.By using the Cramér-Rao inequality, we can prove that $e_s \cdot F(A,s)^{-1} e_s/N$ is the variance of the linear estimates s_N^{L} in the e_s direction of the Bloch sphere. When N is sufficiently large, most of the distribution of linear estimates is included in D_s and the effect of the boundary becomes negligible. Roughly speaking, this condition is represented as $e_s \cdot F(A,s)^{-1} e_s/N \ll (1-\|s\|_2)^2$, where the right hand side is the squared Euclidean distance between s and D_s. This can be rewritten as

$$N \gg \frac{e_s \cdot F(A,s)^{-1} e_s}{(1-\|s\|_2)^2} = \frac{1}{2}N^*. \tag{5.89}$$

We interpret N^* as a reasonable benchmark for judging whether most of the distribution of the linear estimates is included in the physical region or not. The factor of 2 in Eq. (5.88) comes from the Gaussian integration, though in defining N^* it is fairly arbitrary as it makes precise what we mean by 'most' in the preceding sentence. Thus, in order to justify the use of the Cramér-Rao bound for evaluating the estimation error, the number of measurement trials, N, must be larger than N^*.

When $\|s\|_2 < 1$, in the limit of $N \to \infty$, $\text{erfc}[\sqrt{N/N^*}]$ decreases exponentially. This can be readily shown by using the asymptotic expansion [29],

$$\mathrm{erfc}[a] \sim \frac{e^{-a^2}}{\sqrt{\pi}a}\left(1+\sum_{m=1}^{\infty}(-1)^m \frac{1\cdot 3\cdots(2m-1)}{(2a^2)^m}\right). \tag{5.90}$$

Therefore we can see that the approximate expected squared Hilbert-Schmidt distance converges to the Cramér-Rao bound. On the other hand, when $\|s\|_2 = 1$, the second and third terms in Eq. (5.86) disappear and we obtain

$$\bar{\Delta}_N^{\mathrm{HS2}}(\tilde{s}^{\mathrm{ML}}|s) = \frac{1}{4}\frac{1}{N}\left(\mathrm{tr}[F(A,s)^{-1}] - \frac{1}{2}\frac{e_s \cdot F(A,s)^{-2} e_s}{e_s \cdot F(A,s)^{-1} e_s}\right), \tag{5.91}$$

where we assumed that $F(A,s) < \infty$ for a Bloch vector s with $\|s\|_2 = 1$. This is smaller than the Cramér-Rao bound, and this implies that when the true state is pure, a maximum-likelihood estimator can break the Cramér-Rao bound even in the asymptotic region. As explained in Sect. 5.2.3.1, it is proven that the dominant term in the approximate expected loss $\bar{\Delta}_N^{\mathrm{HS2}}(\tilde{s}^{\mathrm{ML}}|s)$ coincides with that in the actual expected loss $\bar{\Delta}_N^{\mathrm{HS2}}(s^{\mathrm{ML}}|s)$. Therefore, we obtain

$$\bar{\Delta}_N^{\mathrm{HS2}}(s^{\mathrm{ML}}|s) = \frac{1}{4N}\left(\mathrm{tr}[F(A,s)^{-1}] - \frac{1}{2}\frac{e_s \cdot F(A,s)^{-2} e_s}{e_s \cdot F(A,s)^{-1} e_s}\right), \tag{5.92}$$

for the Bloch vector s of a pure state. This is the first derivation of the coefficient of the $\frac{1}{N}$-term in $\bar{\Delta}_N^{\mathrm{HS2}}(s^{\mathrm{ML}}|s)$ for pure states.

4. **Expected infidelity**
In order to analyze the expected infidelity, we take the Taylor expansion of the infidelity around the true Bloch vector s up to the second order. Again, using formulas for Gaussian integrals we can derive the approximate expected infidelity. When $\|s\|_2 < 1$,

$$\begin{aligned}\bar{\Delta}_N^{\mathrm{IF}}(\tilde{s}^{\mathrm{ML}}|s) = {} & \frac{1}{4N}\left(\mathrm{tr}[F(A,s)^{-1}] + \frac{s\cdot F(A,s)^{-1}s}{1-(\|s\|_2)^2}\right)\left(1-\frac{1}{2}\mathrm{erfc}\left[\sqrt{\frac{N}{N^*}}\right]\right) \\ & -\frac{1}{4}\frac{1-\|s\|_2}{\sqrt{2\pi e_s\cdot F(A,s)^{-1}e_s}} \\ & \times\left(\mathrm{tr}[F(A,s)^{-1}] - \mathrm{tr}[(Q_s F(A,s) Q_s)^{-}]\right. \\ & \left.+\frac{s\cdot F(A,s)^{-1}s}{1-(\|s\|_2)^2}\right)\frac{e^{-N/N^*}}{\sqrt{N}} \\ & +\frac{1}{4}(1-\|s\|_2)\mathrm{erfc}\left[\sqrt{\frac{N}{N^*}}\right],\end{aligned} \tag{5.93}$$

where

$$Q_s := I - e_s e_s^T \tag{5.94}$$

is the projection matrix onto the subspace orthogonal to $\boldsymbol{s}$, and A^{-} is the Moore-Penrose generalized inverse of a matrix A. From the argument above, we can see that the approximate expected infidelity converges to the Cramér-Rao bound in the limit of large N.

When $\|\boldsymbol{s}\|_2 = 1$, the infidelity is a first order function of $\boldsymbol{s}$, given by $\Delta^{\mathrm{IF}}(\boldsymbol{s}, \boldsymbol{s}') = \frac{1}{2}(1 - \boldsymbol{s} \cdot \boldsymbol{s}')$, and there are no second-order terms. Consequently, the Hesse matrix of the infidelity $H_{\boldsymbol{s}}^{\mathrm{IF}}$ diverges at $\|\boldsymbol{s}\|_2 = 1$. Therefore we cannot apply the Cramér-Rao inequality to the infidelity for pure states. By calculating the expectation value of the approximate estimator $\tilde{\boldsymbol{s}}_N^{\mathrm{ML}}$, we can obtain

$$\bar{\Delta}_N^{\mathrm{IF}}(\tilde{\boldsymbol{s}}^{\mathrm{ML}}|\boldsymbol{s}) = \frac{1}{2}\sqrt{\frac{\boldsymbol{e}_{\boldsymbol{s}} \cdot F(A, \boldsymbol{s})^{-1}\boldsymbol{e}_{\boldsymbol{s}}}{2\pi}}\frac{1}{\sqrt{N}}. \tag{5.95}$$

5.2.3.2 Numerical Analysis

We performed Monte Carlo simulations of one-qubit state tomography using a 6-state POVM $\boldsymbol{\Pi}^{(6)}$.Our task is to estimate the density matrix of the one-qubit system, where the true state can be pure or mixed. We choose a maximum-likelihood estimator, and we used a Newton-Raphson method to solve the (log-)likelihood equation with the completely mixed state $\boldsymbol{s}' = \boldsymbol{0}$ as the initial point of the iteration. When the procedure returned a candidate point outside of the Bloch sphere, we chose the previous point (within the sphere) as the estimate.

The explicit form of $\boldsymbol{\Pi}^{(6)}$ is given in Eq. (5.8). The Fisher matrix and its inverse are given by

$$F(A, \boldsymbol{s}) = \frac{1}{3}\begin{pmatrix} \frac{1}{1-(s_1)^2} & 0 & 0 \\ 0 & \frac{1}{1-(s_2)^2} & 0 \\ 0 & 0 & \frac{1}{1-(s_3)^2} \end{pmatrix},$$

$$F(A, \boldsymbol{s})^{-1} = 3\begin{pmatrix} 1-(s_1)^2 & 0 & 0 \\ 0 & 1-(s_2)^2 & 0 \\ 0 & 0 & 1-(s_3)^2 \end{pmatrix}. \tag{5.96}$$

In Figs. 5.3, 5.4 and 5.5, we show the plots for two loss functions: the squared Hilbert-Schmidt distance Δ^{HS2} and the infidelity Δ^{IF}. The pointwise expected losses $\bar{\Delta}_N(\boldsymbol{s}^{\mathrm{ML}}|\boldsymbol{s})$ and the approximated functions $\bar{\Delta}_N(\tilde{\boldsymbol{s}}^{\mathrm{ML}}|\boldsymbol{s})$ introduced in Sect. 5.2.3.1 are compared, and the accuracy of those approximations are discussed. Table 5.2 is a list of true Bloch vectors $\boldsymbol{s}$ for the figures shown in the following subsections, along with the numerical values of N^* for each $\boldsymbol{s}$. We chose two Bloch radii, $r := \|\boldsymbol{s}\|_2 = 0.9, 0.99$, and two sets of angles $(\theta, \phi) = (0, 0), (\pi/4, \pi/4)$ as the true Bloch vector $\boldsymbol{s}$. For a fixed r, the case with angles $(0, 0)$ corresponds to one of the

Table 5.2 List of the true Bloch vectors under consideration (in spherical coordinates), and numerical values of N^* (rounded down, when possible)

(r, θ, ϕ)	$(0.9, 0, 0)$	$(0.9, \pi/4, \pi/4)$	$(0.99, 0, 0)$	$(0.99, \pi/4, \pi/4)$	$(1, \pi/4, \pi/4)$
Panels	(EIF-1)	(EIF-2)	(EIF-3)	(EHS-1), (EIF-4)	(EHS-2), Fig. 5.5.
N^*	114	417	1194	37947	∞

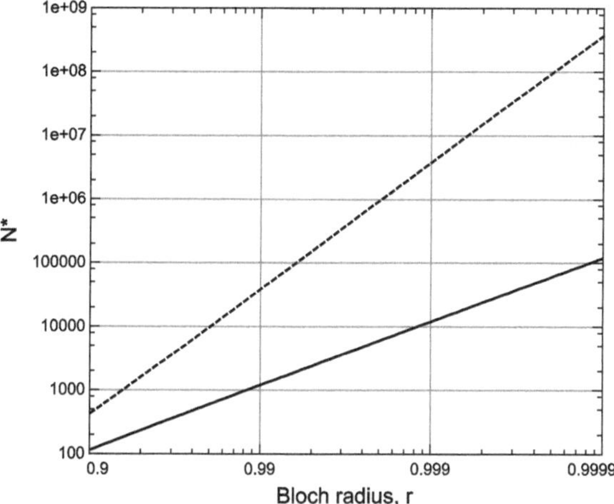

Fig. 5.2 Bloch radius dependency of N^* for standard quantum state tomography, given in Eq. (5.97). The *solid line* is for states s given by $(r, 0, 0)$, and the *dashed line* is for those given by $(r, \pi/4, \pi/4)$. (Reproduced from Ref. [56] with permission. © IOP Publishing Ltd and Deutsche Physikalische Gesellschaft. Published under a CC BY-NC-SA licence)

best case scenarios because the Bloch vector is along one measurement axis, while the $(\pi/4, \pi/4)$ case corresponds to a worst case scenario because the Bloch vector is equidistant from all the measurement axes. The explicit form of N^* for the Fisher matrix in Eq. (5.96) is

$$N^* = 6\left(\frac{1 + \|s\|_2}{1 - \|s\|_2} + 2\frac{(s_1 s_2)^2 + (s_2 s_3)^2 + (s_3 s_1)^2}{(\|s\|_2)^2(1 - \|s\|_2)^2}\right). \tag{5.97}$$

There are two terms which contribute to the divergence at $\|s\|_2 = 1$, and near this value the first term behaves as $O((1 - \|s\|_2)^{-1})$, while the second does so as $O((1 - \|s\|_2)^{-2})$. When the true Bloch vector is along one of the measurement axes, the second term in Eq. (5.97) disappears. For example, if $s = (r, 0, 0)$, we obtain $N^* = 6\frac{1+r}{1-r} \sim \frac{12}{1-r}$ as $r \to 1$. On the other hand, when the true Bloch vector does not lie along any measurement axis, the second term remains. For example, if $s = (r, \pi/4, \pi/4)$, we obtain $N^* = 6\left(\frac{1+r}{1-r} + \frac{5}{8}\frac{1}{r^2(1-r)^2}\right) \sim \frac{15}{4}\frac{1}{(1-r)^2}$. Figure 5.2 is a plot of N^* against r. Therefore N^* for a true Bloch vector whose direction is along one of the measurement axes becomes smaller than that for a true Bloch vector

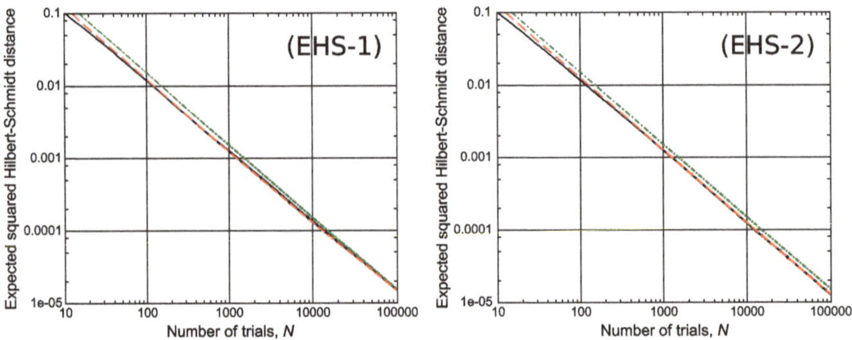

Fig. 5.3 Pointwise expected squared Hilbert-Schmidt distance $\bar{\Delta}_N^{\mathrm{HS2}}$ plotted against the number of measurement trials N: (EHS-1) and (EHS-2) are for the true Bloch vector $\boldsymbol{s}$ given by $(r, \theta, \phi) = (0.99, \pi/4, \pi/4)$ and $(1, \pi/4, \pi/4)$, respectively. The line styles are as follows: *solid (black) line* for the numerically simulated one $\bar{\Delta}_N^{\mathrm{HS2}}(\boldsymbol{s}^{\mathrm{ML}}|\boldsymbol{s})$, *dashed (red) line* for the approximate one $\bar{\Delta}_N^{\mathrm{HS2}}(\tilde{\boldsymbol{s}}^{\mathrm{ML}}|\boldsymbol{s})$ given in Eqs. (5.86) and (5.91), *chain (green) line* for the Cramér-Rao bound, and *dotted (black) vertical line* for N^*. The number of sequences used for the calculation of the statistical expectation values is 10,000. (Reproduced from Ref. [56] with permission.)

whose direction is not. This difference caused by the alignment of measurement axes becomes larger as the purity of $\hat{\rho}(\boldsymbol{s})$ becomes higher.

The terms caused by the bias in Eqs. (5.86) and (5.93) start to decrease exponentially fast after N becomes larger than N^*. We expect that the simulated and approximated plots start to converge to the Cramér-Rao bound after N becomes larger than N^*. In all figures in Sect. 5.2.3, the line styles are as follows: a solid (black) line for the numerically simulated expected loss $\bar{\Delta}_N(\boldsymbol{s}^{\mathrm{ML}}|\boldsymbol{s})$, a dashed (red) line for the approximate expected loss $\bar{\Delta}_N(\tilde{\boldsymbol{s}}^{\mathrm{ML}}|\boldsymbol{s})$ given in Eqs. (5.86), (5.91), (5.93) and (5.95), a chain (green) line for the Cramér-Rao bound, and a dotted (black) vertical line for N^*.

1. **Expected squared Hilbert-Schmidt distance**
   The Cramér-Rao bound of the expected squared Hilbert-Schmidt distance is given by

$$\frac{\mathrm{Tr}\,[H_{\boldsymbol{s}}^{\mathrm{HS2}} F(A, \boldsymbol{s})^{-1}]}{N} = \frac{3}{4}\{3 - (\|\boldsymbol{s}\|_2)^2\}\frac{1}{N}. \tag{5.98}$$

   Figure 5.3 shows the pointwise expected squared Hilbert-Schmidt distance $\bar{\Delta}_N^{\mathrm{HS2}}$ plotted against the number of trials N (the horizontal and vertical axes are both logarithmic scale). The panels (EHS-1) and (EHS-2) are for the true Bloch vector $\boldsymbol{s}$ given by $(r, \theta, \phi) = (0.99, \pi/4, \pi/4)$ and $(r, \theta, \phi) = (1, \pi/4, \pi/4)$, respectively, so that the former is (slightly) mixed, while the latter is pure. The panel (EHS-1) shows that our approximation in Eq. (5.86) converges to the simulated plot, and both the simulated and approximated plots converge to the Cramér-Rao bound of Eq. (5.98) as N becomes large. The same behavior is observed for other mixed true states. On the other hand, panel (EHS-2) shows a different behavior; our approximation in Eq. (5.91) converges to the simulated plot, but the simulated and approximated plots do not converge to the Cramér-Rao bound. This indicates that for pure states, our approximation better captures the behavior of the expected loss than does the Cramér-Rao bound. As mentioned around Eq. (5.91), the reason for this is that the center of the distribution of the linear estimates for a pure state will always be on the boundary of the Bloch sphere, so that about a half of the distribution will always be in the unphysical region. This prohibits a maximum-likelihood estimator from ever converging to the Cramér-Rao bound.

2. **Expected infidelity**
   The infidelity is a nonlinear function of the states, and we must approximate the Cramér-Rao bound in this case; doing so up to second order gives

$$\frac{\mathrm{Tr}\,[H_s^{\mathrm{IF}} F(A, s)^{-1}]}{N} = \frac{3}{4}\left(3 + 2\frac{(s_1 s_2)^2 + (s_2 s_3)^2 + (s_3 s_1)^2}{1 - (\|s\|_2)^2}\right)\frac{1}{N}. \quad (5.99)$$

Figure 5.4 shows the pointwise expected infidelity $\bar{\Delta}_N^{\mathrm{IF}}$ plotted against the number of measurement trials N: (EIF-1), (EIF-2), (EIF-3), (EIF-4) are for the true Bloch vector s given by $(r, \theta, \phi) = (0.9, 0, 0)$, $(0.9, \pi/4, \pi/4)$, $(0.99, 0, 0)$, and $(0.99, \pi/4, \pi/4)$, respectively. Thus panels (EIF-1,2) and panels (EIF-3,4)) are for true states with the same purity. Panels (EIF-1) and (EIF-3) are for the case that one of the measurement axes coincides with the direction of the true Bloch vector, while panels (EIF-2) and (EIF-4) are for the case that all of the measurement axes are as far as possible from the true Bloch vector. Figure 5.4 shows that N^* is a good benchmark for the number of trials required for the simulated plot to start to converge to the Cramér-Rao bound, and so we can say that in order to justify the use of the asymptotic theory, N must be larger than N^*. Figure 5.4 indicates that the angle dependency of the expected infidelity becomes larger as the purity becomes higher. When the true state is far from all measurement axes, the accuracy of our approximation is higher than that of the Cramér-Rao bound. For N smaller than about 10,000 (the 'low N region'), the accuracy of our approximation is low (though still higher than that of the Cramér-Rao bound). We believe that the main reason for our approximation's poor performance in this low N region is the second order approximation of the infidelity, and that higher orders would improve the accuracy here. However, in the high N region the approximation can be seen to capture the behavior of the curve far better than the Cramér-Rao bound.

Figure 5.5 shows the pointwise expected infidelity $\bar{\Delta}_N^{\mathrm{IF}}$ against the number of measurement trials N for the true Bloch vector s given by $(r, \theta, \phi) = (1, \pi/4, \pi/4)$. For pure true states, the expected infidelity decreases as $O(\sqrt{N})$, and Fig. 5.5 shows that the expected infidelity converges to the approximate function.

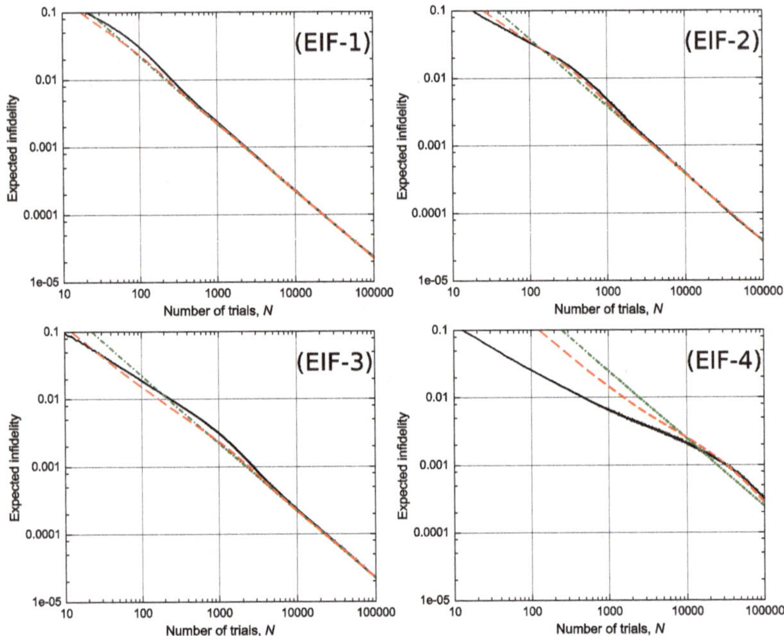

Fig. 5.4 Pointwise expected infidelity $\bar{\Delta}_N^{\mathrm{IF}}$ plotted against the number of measurement trials N: (EIF-1), (EIF-2), (EIF-3), (EIF-4) are for the true Bloch vector $\boldsymbol{s}$ given by $(r, \theta, \phi) = (0.9, 0, 0), (0.9, \pi/4, \pi/4), (0.99, 0, 0)$, and $(0.99, \pi/4, \pi/4)$, respectively. The line styles are as follows: *solid (black) line* for the numerically simulated expected infidelity $\bar{\Delta}_N^{\mathrm{IF}}(\boldsymbol{s}^{\mathrm{ML}}|\boldsymbol{s})$, *dashed (red) line* for the approximate expected infidelity $\bar{\Delta}_N^{\mathrm{IF}}(\tilde{\boldsymbol{s}}^{\mathrm{ML}}|\boldsymbol{s})$ given in Eq. (5.93), *chain (green) line* for the Cramér-Rao bound, and dotted (black) vertical line for N^*. The number of sequences used for the calculation of the statistical expectation values is 10,000. (Reproduced from Ref. [56] with permission. © IOP Publishing Ltd and Deutsche Physikalische Gesellschaft. Published under a CC BY-NC-SA licence)

5.3 Error Probability

In the Sect. 5.2, we explained our results on expected losses of extended linear, extendeed norm-minimization, and maximum-likelihood estimators in quantum tomography with finite data. In this section, we explain our results on error probability for those estimators in the same setting. In Sect. 5.3.1, by using the Hoeffding inequality, we derive functions which upper-bound the error probabilities of an extended linear estimator. In Secs. 5.3.2 and 5.3.3, we focus on two estimators, an extended norm-minimization and maximum-likelihood estimator, which always give physical estimates. We derive functions upper-bounding the error probabilities of these estimators by combining the results of an extended linear estimator explained in Sect. 5.3.1 and some inequalities on norms and loss functions.

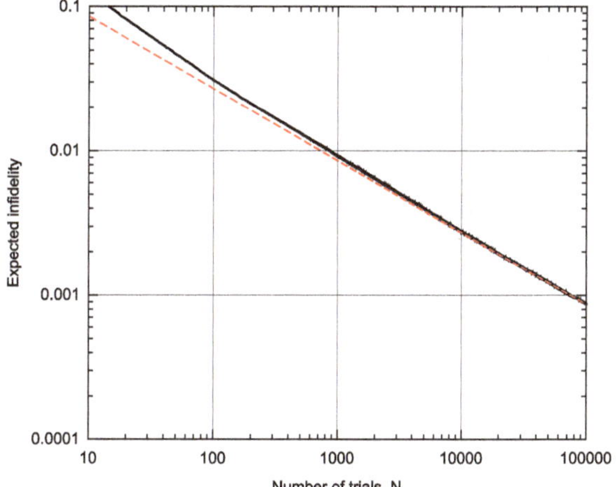

Fig. 5.5 Pointwise expected infidelity $\bar{\Delta}_N^{\mathrm{IF}}$ plotted against the number of measurement trials N for the true Bloch vector $\boldsymbol{s}$ given by $(r, \theta, \phi) = (1, \pi/4, \pi/4)$. the line styles are as follows: *solid (black) line* for the numerically simulated expected infidelity $\bar{\Delta}_N^{\mathrm{IF}}(\boldsymbol{s}^{\mathrm{ML}}|\boldsymbol{s})$, *dashed (red) line* for the approximate expected infidelity $\tilde{\Delta}_N^{\mathrm{IF}}(\tilde{\boldsymbol{s}}^{\mathrm{ML}}|\boldsymbol{s})$ given in Eq. (5.95). The number of sequences used for the calculation of statistical expectation values is 10,000. (Reproduced from Ref. [56] with permission. © IOP Publishing Ltd and Deutsche Physikalische Gesellschaft. Published under a CC BY-NC-SA licence)

5.3.1 Extended Linear Estimator

First, we derive a function which upper-bounds the error probability for an extended linear estimator in quantum state tomography.

Let us choose the ℓ_∞-distance as a loss function on the parameter space. Then we have the following theorem.

Theorem 5.5 (Error probability, $\boldsymbol{s}^{\mathrm{eL}}$, ℓ_∞-distance)
When we choose the ℓ_∞-distance as the loss function on the generalized Bloch vector, we have the following upper bound of the error probability for an extended linear estimator.

$$\mathrm{P}_{\delta,N}^{\ell_\infty}(A, \boldsymbol{s}^{\mathrm{eL}}|\boldsymbol{s}) \le 2\sum_{\alpha=1}^{d^2-1} \exp\left[-2N\delta^2/c_\alpha\right], \ \forall \boldsymbol{s} \in B, \tag{5.100}$$

where

$$c_\alpha := \sum_{j=1}^{J} r^{(j)} \left\{\max_m (\Lambda_{\mathrm{left}}^{-1})_{\alpha,(j,m)} - \min_m (\Lambda_{\mathrm{left}}^{-1})_{\alpha,(j,m)}\right\}^2, \ \alpha = 1, \dots, d^2 - 1. \tag{5.101}$$

Proof (Theorem 5.5)
The statement $\|\boldsymbol{s}_N^{\mathrm{eL}} - \boldsymbol{s}\|_\infty > \delta$ implies that there exists at least one α such that $|s_{N,\alpha}^{\mathrm{eL}} - s_\alpha| > \delta$. This implies

$$\left\{\boldsymbol{x}^N \mid \|\boldsymbol{s}_N^{\mathrm{eL}} - \boldsymbol{s}\|_\infty > \delta\right\} \subseteq \bigcup_{\alpha=1}^{d^2-1} \left\{\boldsymbol{x}^N \mid |s_{N,\alpha}^{\mathrm{eL}} - s_\alpha| > \delta\right\}, \tag{5.102}$$

and we obtain

$$\mathbb{P}\left[\|\boldsymbol{s}_N^{\mathrm{eL}} - \boldsymbol{s}\|_\infty > \delta\right] \leq \sum_{\alpha=1}^{d^2-1} \mathbb{P}\left[|s_{N,\alpha}^{\mathrm{eL}} - s_\alpha| > \delta\right]. \tag{5.103}$$

Let $m_q^{(j)}$ denote the q-th outcome in $n^{(j)}$ trials of the measurement $\boldsymbol{\Pi}^{(j)}$ ($j = 1, \ldots, J,\ q = 1, \ldots, n^{(j)}$). From Eq. (5.32), we have

$$s_{N,\alpha}^{\mathrm{eL}} - s_\alpha = \sum_{j=1}^{J} \sum_{m=1}^{M^{(j)}} (\Lambda_{\mathrm{left}}^{-1})_{\alpha,(j,m)} \left\{\nu^{(j)}(m) - p^{(j)}(m)\right\} \tag{5.104}$$

$$= \sum_{j=1}^{J} \sum_{q=1}^{n^{(j)}} \left\{ \frac{(\Lambda_{\mathrm{left}}^{-1})_{\alpha,(j,m_q^{(j)})}}{n^{(j)}} - \sum_{m=1}^{M^{(J)}} \frac{(\Lambda_{\mathrm{left}}^{-1})_{\alpha,(j,m)}}{n^{(j)}} p^{(j)}(m) \right\} \tag{5.105}$$

Each random variable in the sum above takes value in

$$\left[\frac{1}{n^{(j)}} \min_m (\Lambda_{\mathrm{left}}^{-1})_{\alpha,(j,m)}, \frac{1}{n^{(j)}} \max_m (\Lambda_{\mathrm{left}}^{-1})_{\alpha,(j,m)}\right]. \tag{5.106}$$

By applying Hoeffding's tail inequality (Eq. (3.38)), we obtain

$$\mathbb{P}[|s_{N,\alpha}^{\mathrm{eL}} - s_\alpha| > \delta] \leq 2\exp\left[-2N\delta^2/c_\alpha\right]. \tag{5.107}$$

Therefore we obtain

$$\mathbb{P}[\|\boldsymbol{s}_N^{\mathrm{eL}} - \boldsymbol{s}\|_\infty > \delta] \leq 2\sum_{\alpha=1}^{d^2-1} \exp\left[-2N\delta^2/c_\alpha\right]. \tag{5.108}$$

□

Let us consider different loss functions, the ℓ_1- and ℓ_2-distances. By combining Theorem 5.5 and some norm inequalities, we obtain the following theorem.

Theorem 5.6 (Error probability, s^{eL}, ℓ_1- and ℓ_2-distances)
When we choose the ℓ_1-distance or ℓ_2-distance as the loss function for the generalized Bloch vector, we have the following upper bounds on the error probabilities for an

extended linear estimator.

$$\mathrm{P}^{\ell_1}_{\delta,N}(A, s^{\mathrm{eL}}|s) \le 2\sum_{\alpha=1}^{d^2-1} \exp\left[-\frac{2}{(d^2-1)^2}\frac{\delta^2}{c_\alpha}N\right], \tag{5.109}$$

$$\mathrm{P}^{\ell_2}_{\delta,N}(A, s^{\mathrm{eL}}|s) \le 2\sum_{\alpha=1}^{d^2-1} \exp\left[-\frac{2}{d^2-1}\frac{\delta^2}{c_\alpha}N\right], \tag{5.110}$$

for any $s \in B_d$.

Proof (Theorem 5.6)
From Eqs. (5.163) and (5.164), we can prove that

$$\|s^{\mathrm{eL}}_N - s\|_1 > \delta \Longrightarrow \|s^{\mathrm{eL}}_N - s\|_\infty > \frac{\delta}{d^2-1}, \tag{5.111}$$

and

$$\|s^{\mathrm{eL}}_N - s\|_2 > \delta \Longrightarrow \|s^{\mathrm{eL}}_N - s\|_\infty > \frac{\delta}{\sqrt{d^2-1}}. \tag{5.112}$$

Therefore we have

$$\mathbb{P}\left[\ \|s^{\mathrm{eL}}_N - s\|_1 > \delta\right] \le \mathbb{P}\left[\ \|s^{\mathrm{eL}}_N - s\|_\infty > \frac{\delta}{d^2-1}\right] \tag{5.113}$$

$$\le 2\sum_{\alpha=1}^{d^2-1} \exp\left[-\frac{2}{(d^2-1)^2}\frac{\delta^2}{c_\alpha}N\right] \tag{5.114}$$

and

$$\mathbb{P}\left[\ \|s^{\mathrm{eL}}_N - s\|_2 > \delta\right] \le \mathbb{P}\left[\ \|s^{\mathrm{eL}}_N - s\|_\infty > \frac{\delta}{\sqrt{d^2-1}}\right] \tag{5.115}$$

$$\le 2\sum_{\alpha=1}^{d^2-1} \exp\left[-\frac{2}{d^2-1}\frac{\delta^2}{c_\alpha}N\right]. \tag{5.116}$$

□

The difference between Eqs. (5.109) and (5.110) is the rate of decrease of their R.H.S., namely $\frac{\delta^2}{(d^2-1)^2 c_\alpha}$ and $\frac{\delta^2}{(d^2-1)c_\alpha}$, respectively. Unlike in Eq. (5.100), these are dependent of the dimension of the system and become smaller as d becomes larger. We note that these are probably not tight upper bounds on the error probabilities, and may not be an essential property of ℓ_1- and ℓ_2-distances.

From Eqs. (5.66) and (5.67), we obtain

$$\mathrm{P}^{\mathrm{HS}}_{\delta,N}(\breve{\boldsymbol{\Pi}}, \hat{\rho}^{\mathrm{est}}|\hat{\rho}) = \mathrm{P}^{\ell_2}_{2\delta,N}(A, \boldsymbol{s}^{\mathrm{est}}|\boldsymbol{s}), \tag{5.117}$$

$$\mathrm{P}^{\mathrm{T}}_{\delta,N}(\breve{\boldsymbol{\Pi}}, \hat{\rho}^{\mathrm{est}}|\hat{\rho}) \le \mathrm{P}^{\mathrm{HS}}_{\sqrt{\frac{2}{d}}\delta,N}(\breve{\boldsymbol{\Pi}}, \hat{\rho}^{\mathrm{est}}|\hat{\rho}) = \mathrm{P}^{\ell_2}_{2\sqrt{\frac{2}{d}}\delta,N}(A, \boldsymbol{s}^{\mathrm{est}}|\boldsymbol{s}). \tag{5.118}$$

Therefore we have the following theorem.

Theorem 5.7 (Error probability, $\hat{\rho}^{\mathrm{eL}}$, Δ^{HS}, Δ^{T})
When we choose the Hilbert-Schmidt distance or Trace distance as the loss function for the density matrix, we have the following upper bounds on the error probabilities for an extended linear estimator.

$$\mathrm{P}^{\mathrm{HS}}_{\delta,N}(\breve{\boldsymbol{\Pi}}, \hat{\rho}^{\mathrm{eL}}|\hat{\rho}) \le 2\sum_{\alpha=1}^{d^2-1} \exp\left[-\frac{8}{d^2-1}\frac{\delta^2}{c_\alpha}N\right], \tag{5.119}$$

$$\mathrm{P}^{\mathrm{T}}_{\delta,N}(\breve{\boldsymbol{\Pi}}, \hat{\rho}^{\mathrm{eL}}|\hat{\rho}) \le 2\sum_{\alpha=1}^{d^2-1} \exp\left[-\frac{16}{d(d^2-1)}\frac{\delta^2}{c_\alpha}N\right], \tag{5.120}$$

for any true density matrix $\hat{\rho}$.

5.3.2 Extended Norm-Minimization Estimator

In this subsection, we derive functions that upper-bound error probabilities of an extended norm-minimization estimator in quantum tomography.

Theorem 5.8 (Error probability, $\boldsymbol{s}^{\ell_2\text{-eNM}}$, ℓ_1-ℓ_2-ℓ_∞-distances)
When we choose the ℓ_1-distance or ℓ_2-distance as the loss function, for the extended norm-minimization estimator with respect to ℓ_2-norm, we have

$$\mathrm{P}^{\ell_1}_{\delta,N}(A, \boldsymbol{s}^{\ell_2\text{-eNM}}|\boldsymbol{s}) \le 2\sum_{\alpha=1}^{d^2-1} \exp\left[-\frac{2}{(d^2-1)^2}\frac{\delta^2}{c_\alpha}N\right], \tag{5.121}$$

$$\mathrm{P}^{\ell_2}_{\delta,N}(A, \boldsymbol{s}^{\ell_2\text{-eNM}}|\boldsymbol{s}) \le 2\sum_{\alpha=1}^{d^2-1} \exp\left[-\frac{2}{d^2-1}\frac{\delta^2}{c_\alpha}N\right], \tag{5.122}$$

$$\mathrm{P}^{\ell_\infty}_{\delta,N}(A, \boldsymbol{s}^{\ell_2\text{-eNM}}|\boldsymbol{s}) \le 2\sum_{\alpha=1}^{d^2-1} \exp\left[-\frac{2}{d^2-1}\frac{\delta^2}{c_\alpha}N\right]. \tag{5.123}$$

Proof (Theorem 5.8)
From Eq. (5.70), we can prove that

$$\|\boldsymbol{s}^{\ell_2\text{-eNM}}_N - \boldsymbol{s}\|_2 > \delta \Longrightarrow \|\boldsymbol{s}^{\mathrm{eL}}_N - \boldsymbol{s}\|_2 > \delta. \tag{5.124}$$

Therefore

$$\left\{ x^N | \ \|s_N^{\ell_2\text{-eNM}} - s\|_2 > \delta \right\} \subseteq \left\{ x^N | \ \|s_N^{\text{eL}} - s\|_2 > \delta \right\} \tag{5.125}$$

holds, and we obtain

$$\mathbb{P}\left[\|s_N^{\ell_2\text{-eNM}} - s\|_2 > \delta \right] \leq \mathbb{P}\left[\|s_N^{\text{eL}} - s\|_2 > \delta \right]. \tag{5.126}$$

By combining Eqs. (5.126) and (5.110), we obtain Eq. (5.122). Equations (5.121) and (5.123) are obtained by combining Eq. (5.122) with Eqs. (5.162) and (5.164), respectively. □

By combining Eqs. (5.117), (5.118), and (5.122), we obtain the following theorem.

Theorem 5.9 (Error probability, $\hat{\rho}^{\ell_2\text{-eNM}}$, Δ^{HS}, Δ^{T}, Δ^{IF}) When we choose the Hilbert-Schmidt distance, Trace distance, or infidelity as the loss function for the density matrix, we have the following upper bounds on the error probabilities for the extended ℓ_2-norm-minimization estimator.

$$\mathrm{P}_{\delta,N}^{\text{HS}}(\breve{\boldsymbol{\Pi}}, \hat{\rho}^{\ell_2\text{-eNM}} | \hat{\rho}) \leq 2 \sum_{\alpha=1}^{d^2-1} \exp\left[-\frac{8}{d^2-1} \frac{\delta^2}{c_\alpha} N \right], \tag{5.127}$$

$$\mathrm{P}_{\delta,N}^{\text{T}}(\breve{\boldsymbol{\Pi}}, \hat{\rho}^{\ell_2\text{-eNM}} | \hat{\rho}) \leq 2 \sum_{\alpha=1}^{d^2-1} \exp\left[-\frac{16}{d(d^2-1)} \frac{\delta^2}{c_\alpha} N \right], \tag{5.128}$$

$$\mathrm{P}_{\delta,N}^{\text{IF}}(\breve{\boldsymbol{\Pi}}, \hat{\rho}^{\ell_2\text{-eNM}} | \hat{\rho}) \leq 2 \sum_{\alpha=1}^{d^2-1} \exp\left[-\frac{4}{d(d^2-1)} \frac{\delta^2}{c_\alpha} N \right], \tag{5.129}$$

for any true density matrix $\hat{\rho}$.

Let us consider quantum state tomography of a k-qubit system and suppose that we make the three Pauli measurements with detection efficiency η on each qubit. There are 3^k different tensor products of Pauli matrices ($J = 3^k$), and suppose that we observe each equally $n := N/3^k$ times. Let us choose $\boldsymbol{\sigma}$ to be the set of tensor products of Pauli and identity matrices with the normalization factor $1/\sqrt{2^{k-1}}$. From the relation $s_\alpha = \text{Tr}[\rho\hat{\sigma}_\alpha]$, we obtain $c_\alpha = 2^{3-k} \cdot 3^{k-l}/\eta^{2(k-l)}$, where $l = 0, \ldots, k-1$ is the number of the identity matrices appearing in $\hat{\sigma}_\alpha$. From the information above, we can derive the explicit form of P_u^Δ for k-qubit state tomography. When we choose the trace distance as the loss function, we have

$$\mathrm{P}_u^{\text{T}}(\delta, N, \breve{\boldsymbol{\Pi}}, \hat{\rho}^{\ell_2\text{-eNM}}) = 2 \sum_{l=0}^{k-1} 3^{k-l} \binom{k}{l} \exp\left[-\frac{2}{2^{2k}-1} \frac{\eta^{2(k-l)}}{3^{k-l}} \delta^2 N \right]. \tag{5.130}$$

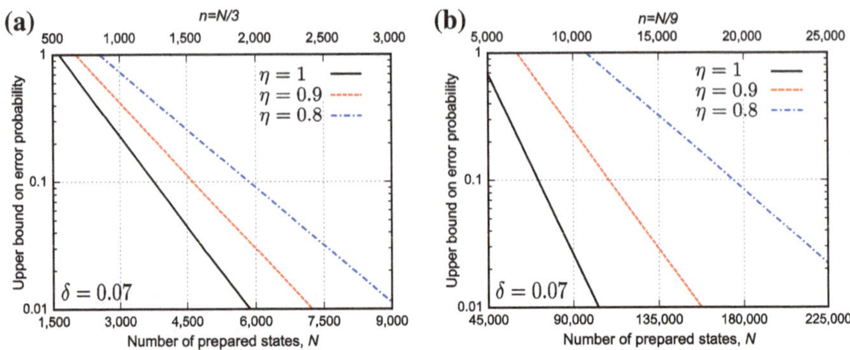

Fig. 5.6 Upper bound on error probability of $\hat{\rho}^{\ell_2\text{-eNM}}$ for error threshold $\delta = 0.07$ in quantum state tomography: panel **a** is the one-qubit case ($k = 1$) and panel **b** is the two-qubit case ($k = 2$). In both panels, the vertical axes are $\mathrm{P}_u^{\mathrm{T}}$ in Eq. (5.130). The *lower* and *upper* horizontal axes are the number of prepared states N and the number of observations for each tensor product of Pauli matrices $n = N/3^k$, respectively. In both panels, the line styles are fixed as follows: *solid* (*black*) *line* for detection efficiency $\eta = 1$, *dashed* (*red*) *line* for $\eta = 0.9$, *chain* (*blue*) *line* for $\eta = 0.8$. (Reproduced from Ref. [55] with permission)

The details of the derivation of Eq. (5.130) are explained in B.1. Figure 5.6 shows plots of Eq. (5.130) for the one-qubit ($k = 1$) and two-qubit ($k = 2$) cases in panels (a) and (b), respectively. The error threshold is $\delta = 0.07$ and detection efficiency is $\eta = 1,\ 0.9,\ 0.8$. Both panels indicate that smaller detection efficiency requires a larger number of prepared states. The plots tell us what value of N sufficient for guaranteeing a fixed confidence level. For example, if we want to guarantee 99 % confidence level for $\delta = 0.07$ in one-qubit state tomography with $\eta = 0.9$, panel (a) indicates that $N = 7500$ is sufficient for that.

In [30], an efficient maximum-likelihood (ML) estimator algorithm is proposed for quantum state tomography using an IC set of projective measurements with Gaussian noise whose variance is known, and numerical results for k-qubit ($k = 1, \ldots, 9$) state tomography indicate that the computational cost would be significantly lower than that of standard ML algorithms. In general, a ML estimator is different from the ℓ_2-eNM estimator, but in the setting considered in [30] the ML estimator is a specific case of the ℓ_2-eNM estimator, which is defined for general IC measurements. Despite this generality, we find that their efficient algorithm can be modified and used for our ℓ_2-eNM estimates.[1] Additionally, our result (Theorem 5.9) shows that the ℓ_2-eNM estimator can be used without assuming projective measurements or Gaussian noise.

It is natural to ask if instead of performing two sequential optimizations as in the ℓ_2-eNM case one performs a single constrained optimization. This is well-known in classical statistics and was applied to a quantum estimation problem in [31]. Define

[1] These calculations involve numerical errors, and strictly speaking the calculated estimate is different from the exact ℓ_2-eNM estimate. We analyze the effect of numerical (and systematic) errors on error thresholds and upper bounds on error probability and give a solution to this problem in Sects. B.2 and B.3.

a constrained least squares estimator

$$\hat{\rho}_N^{\mathrm{CLS}} := \operatorname*{argmin}_{\hat{\rho}' \in \mathcal{S}(\mathcal{H})} \|\boldsymbol{p}(\hat{\rho}') - \boldsymbol{\nu}_N\|_2, \tag{5.131}$$

which always exists and is always physical. Using nearly the same proof as in Theorem 5.9, we can derive P_u^{Δ} for $\hat{\rho}^{\mathrm{CLS}}$. The result is equivalent or larger than Theorem 5.9—the details and a comparison to ρ^{ENM} are shown in Sect. B.4. Although in some cases the upper-bounds on their error probabilities coincide, in order to calculate CLS estimates we need to solve the quadratic optimization problem under inequality constraints, which the ℓ_2-eNM case avoids.

5.3.3 *Maximum-Likelihood Estimator*

In this subsection, we present our results on error probability for a maximum-likelihood estimator. Suppose that we observe each POVM equally, i.e., $n^{(j)} = n := N/J$ times. Let us introduce $M := \sum_{j=1}^{J} M^{(j)}$ and define $\|\Lambda\|_{1,\min}$ and $\|\Lambda\|_{1,\max}$ as

$$\|\Lambda\|_{1,\min} := \min_{\boldsymbol{v}\neq \boldsymbol{0}} \frac{\|\Lambda \boldsymbol{v}\|_1}{\|\boldsymbol{v}\|_1}. \tag{5.132}$$

First, let us consider two loss functions, the ℓ_1- and ℓ_2-distances.

Theorem 5.10 (Error probability, s^{ML}, ℓ_1-,ℓ_2-distances)
When we choose the ℓ_1-distance or ℓ_2-distance as the loss function, for the maximum-likelihood estimator, we have

$$\mathrm{P}_{\delta,N}^{\ell_1}(A, \boldsymbol{s}^{\mathrm{ML}}|\boldsymbol{s}) \le \frac{(n+M-1)!}{n!(M-1)!} \exp\left[-\frac{1}{8}(\|\Lambda\|_{1,\min})^2 \delta^2 n\right], \tag{5.133}$$

$$\mathrm{P}_{\delta,N}^{\ell_2}(A, \boldsymbol{s}^{\mathrm{ML}}|\boldsymbol{s}) \le \frac{(n+M-1)!}{n!(M-1)!} \exp\left[-\frac{1}{8}(\|\Lambda\|_{1,\min})^2 \delta^2 n\right]. \tag{5.134}$$

Proof (Theorem 5.10)
From Eq. (5.29) and the definition of $\|\Lambda\|_{1,\min}$, we obtain

$$\|\boldsymbol{p}(\hat{\rho}_N^{\mathrm{ML}}) - \boldsymbol{p}(\hat{\rho})\|_1 = \|\Lambda(\boldsymbol{s}_N^{\mathrm{ML}} - \boldsymbol{s})\|_1 \tag{5.135}$$

$$\ge \|\Lambda\|_{1,\min} \cdot \|\boldsymbol{s}_N^{\mathrm{ML}} - \boldsymbol{s}\|_1. \tag{5.136}$$

On the other hands, from the triangle inequality for the ℓ_1-norm, we have

$$\|\boldsymbol{p}(\hat{\rho}_N^{\mathrm{ML}}) - \boldsymbol{p}(\hat{\rho})\|_1 \le \|\boldsymbol{p}(\hat{\rho}_N^{\mathrm{ML}}) - \boldsymbol{\nu}_N\|_1 + \|\boldsymbol{\nu}_N - \boldsymbol{p}(\hat{\rho})\|_1. \tag{5.137}$$

From Pinsker's inequality (Eq. (5.165)), we obtain

$$\|\boldsymbol{p}(\hat{\rho}_N^{\mathrm{ML}}) - \boldsymbol{\nu}_N\|_1 \leq \sqrt{2K(\boldsymbol{\nu}_N \| \boldsymbol{p}(\hat{\rho}_N^{\mathrm{ML}}))} \leq \sqrt{2K(\boldsymbol{\nu}_N \| \boldsymbol{p}(\hat{\rho}))}, \tag{5.138}$$

$$\|\boldsymbol{\nu}_N - \boldsymbol{p}(\hat{\rho})\|_1 \leq \sqrt{2K(\boldsymbol{\nu}_N \| \boldsymbol{p}(\hat{\rho}))}. \tag{5.139}$$

Therefore

$$\begin{aligned} \|\Lambda\|_{1,\min} \cdot \|\boldsymbol{s}_N^{\mathrm{ML}} - \boldsymbol{s}\|_1 &\leq \|\boldsymbol{p}(\hat{\rho}_N^{\mathrm{ML}}) - \boldsymbol{p}(\hat{\rho})\|_1 \\ &\leq 2\sqrt{2K(\boldsymbol{\nu}_N \| \boldsymbol{p}(\hat{\rho}))} \end{aligned} \tag{5.140}$$

holds for any $\boldsymbol{\nu}_N$ and $\boldsymbol{\rho}$ $(\boldsymbol{s})$, and we obtain

$$\mathbb{P}\left[\|\boldsymbol{s}_N^{\mathrm{ML}} - \boldsymbol{s}\|_1 > \delta\right] \leq \mathbb{P}\left[K(\boldsymbol{\nu}_N \| \boldsymbol{p}(\hat{\rho})) > (\|\Lambda\|_{1,\min} \cdot \delta)^2/8\right]. \tag{5.141}$$

From the improved Sanov's inequality (Eq. (3.41)), we have

$$\begin{aligned} &\mathbb{P}\left[K(\boldsymbol{\nu}_N \| \boldsymbol{p}(\hat{\rho})) > (\|\Lambda\|_{1,\min} \cdot \delta)^2/8\right] \\ &\leq \frac{(n+M-1)!}{n!(M-1)!} \exp\left[-n \cdot \inf_{\boldsymbol{q}: K(\boldsymbol{q} \| \boldsymbol{p}(\hat{\rho})) > (\|\Lambda\|_{1,\min} \cdot \delta)^2/8} K(\boldsymbol{q} \| \boldsymbol{p}(\hat{\rho}))\right] \quad (5.142) \\ &= \frac{(n+M-1)!}{n!(M-1)!} \exp\left[-(\|\Lambda\|_{1,\min})^2 \delta^2 n/8\right], \quad (5.143) \end{aligned}$$

and we obtain Eq. (5.133). Equation (5.134) can be derived by combining Eqs. (5.133) and (5.162). □

By combining Eqs. (5.117), (5.118), and (5.122) with Theorem 5.10. we obtain the following theorem.

Theorem 5.11 (Error probability, $\hat{\rho}^{\mathrm{ML}}$, Δ^{HS}, Δ^{T}, Δ^{IF})
When we choose the Hilbert-Schmidt distance, Trace distance, or infidelity as the loss function for the density matrix, we have the following upper bounds on the error probabilities for the maximum-likelihood estimator.

$$\mathrm{P}_{\delta,N}^{\mathrm{HS}}(\breve{\boldsymbol{\Pi}}, \hat{\rho}^{\mathrm{ML}} | \hat{\rho}) \leq \frac{(n+M-1)!}{n!(M-1)!} \exp\left[-\frac{1}{2}(\|\Lambda\|_{1,\min})^2 \delta^2 n\right], \tag{5.144}$$

$$\mathrm{P}_{\delta,N}^{\mathrm{T}}(\breve{\boldsymbol{\Pi}}, \hat{\rho}^{\mathrm{ML}} | \hat{\rho}) \leq \frac{(n+M-1)!}{n!(M-1)!} \exp\left[-\frac{1}{d}(\|\Lambda\|_{1,\min})^2 \delta^2 n\right], \tag{5.145}$$

$$\mathrm{P}_{\delta,N}^{\mathrm{IF}}(\breve{\boldsymbol{\Pi}}, \hat{\rho}^{\mathrm{ML}} | \hat{\rho}) \leq \frac{(n+M-1)!}{n!(M-1)!} \exp\left[-\frac{1}{4d}(\|\Lambda\|_{1,\min})^2 \delta^2 n\right], \tag{5.146}$$

for any true density matrix $\hat{\rho}$.

The upper bounds on error probability for $\hat{\rho}^{\mathrm{ML}}$ in Theorem 5.11 is larger than those for $\rho^{\ell_2\text{-eNM}}$ in Theorem 5.9. For example of one-qubit state tomography using three Pauli projective measurements with detection efficency 1, $d = 2$, $J = 3$, $M = 6$, and

$$\Lambda = \frac{1}{2} \begin{pmatrix} 1 & 0 & 0 \\ -1 & 0 & 0 \\ 0 & 1 & 0 \\ 0 & -1 & 0 \\ 0 & 0 & 1 \\ 0 & 0 & -1 \end{pmatrix}. \tag{5.147}$$

Then we have $\|\Lambda\|_{1,\min} = 1/2$ and

$$\mathrm{P}_u^{\mathrm{T}}(\delta, N, \breve{\boldsymbol{\Pi}}, \hat{\rho}^{\mathrm{ML}}) = \frac{(N/3+5)!}{(N/3)!5!} \exp\left[-\frac{1}{24}\delta^2 N\right]. \tag{5.148}$$

Obviously, Eq. (5.148) is larger than Eq. (5.130) for $k = 1$,

$$\mathrm{P}_u^{\mathrm{T}}(\delta, N, \breve{\boldsymbol{\Pi}}, \hat{\rho}^{\ell_2\text{-eNM}}) = 6 \exp\left[-\frac{2}{9}\delta^2 N\right]. \tag{5.149}$$

However, this does not mean we can immediately conclude that $\hat{\rho}^{\mathrm{ML}}$ is less precise than $\hat{\rho}^{\ell_2\text{-eNM}}$ because their upper bounds on error probability are probably not optimal. The author believes that $\hat{\rho}^{\mathrm{ML}}$ is intrinsically as precise as or more precise than $\hat{\rho}^{\ell_2\text{-eNM}}$ and that the techniques used in the proof of Theorem 5.11 is not appropriate. A maximum-likelihood estimator is one of the most popular estimator in current experiments of quantum tomography, and it is important to derive a smaller upper bound for $\hat{\rho}^{\mathrm{ML}}$. This is a topic for future research.

5.4 History and Related Works

The first systematic approach for quantum state estimation was done by Fano in 1957 [16], where a linear estimator for finite-dimensional systems was introduced, but the statistical estimation errors were not discussed. In the 1960s and 1970s, statistical estimation errors in quantum estimation were analyzed, and the results were organized into the textbooks of Helstrom [32] and Holevo [33]. Representative papers for the asymptotic theory of quantum estimation after the eighties are collected in [34]. The main interest of these works was to derive the fundamental lower-bounds of estimation errors originating in quantum mechanics. Our approach is to derive the upper-bounds, and so is complementary to that of conventional quantum estimation theory.

The term "tomography" was first used for identifying a given quantum state by Smithey et al. [1]. The physical system performed tomography in [1] was a spatial-temporal light mode, the measurement apparatus was a Homodyne detector, and the reconstruction scheme was a linear estimator proposed in [35]. The linear inversion used was the inverse Radon transform, as this had been used in medical imaging, where it is called computed tomography (CT) [36]. This is the reason that the identification of a quantum object is called "tomography". After the paper was published, the term quantum tomography started to be used not only for estimations with the Homodyne detector and inverse Radon transform, but for independent experimental design and any estimator [37]. Nowadays, quantum tomography has been applied to many types of physical systems other than spatial-temporal light modes, for example, photons [38], an electron in a semiconductor [39], superconducting qubits [40–42], trapped ions [43], cold atoms [44, 45], single NV centers in diamond [46], and molecules [47–49].

5.5 Summary

In Chap. 5, we explained the estimation setting in quantum tomography and our results on finite sample analysis of expected losses and error probabilities in this setting.

In Sect. 5.1, we categorized quantum tomography into four types by the estimation object; quantum state tomography, quantum process tomography, POVM tomography, and quantum instrument tomography. In order to simplify the explanation, we focused on quantum state tomography. We explained the parametrization, experimental design, estimator, loss function, and figure of merit in quantum tomography. For the figure of merit, we proposed a rigorous method for evaluating estimation errors with finite samples by using functions upper-bounding expected losses and error probability.

In Sect. 5.2, we explained our results on finite sample analysis of expected losses in quantum tomography. We focused on three estimators; extended linear, extended norm-minimization, and maximum-likelihood. In Sect. 5.2.1, we derived an upper bound on the expected loss of an extended linear estimator. In Sect. 5.2.2, we proposed a new estimator called an extended norm-minimization and derived an upper bound on the expected loss of the estimator by using the upper bound of an extended linear estimator derived in Sect. 5.2.1. In Sect. 5.2.3, we analyzed expected losses of a maximum-likelihood estimator in one-qubit state tomography. We derived an explicit formula of the expected squared Hilbert-Schmidt distance and the expected infidelity between a maximum-likelihood estimate and the true state under two approximations: a Gaussian distribution matched to the moments of the asymptotic multinomial distribution, and a linearization of the parameter space boundary imposed by the positivity of quantum states. We performed Monte Carlo simulations of one-qubit state tomography and evaluated the accuracy of the approximation formulas by comparing them to the numerical results. The numerical comparison shows that our approximation reproduces the behavior in the nonasymptotic regime much better than the asymptotic theory, and the typical number of measurement trials derived from the approximation is a reasonable threshold after which the expected loss starts to converge to the asymptotic behavior.

In Sect. 5.3, we explained our results on finite sample analysis of error probabilities in quantum tomography. As in Sect. 5.2, we focused on three estimators; extended linear, extended norm-minimization, and maximum-likelihood estimators. In Sect. 5.3.1, by using the Hoeffding inequality and some norm inequalities, we derived upper bounds on error probabilities for an extended linear estimator. In Sect. 5.3.2, by using the upper bounds obtained in Sect. 5.3.1, we derived upper bounds for the extended ℓ_2-norm minimization estimator. We also gave the explicit form of the upper bounds for k-qubit state tomography and showed the plots for $k = 1$ and $k = 2$ cases. In Sect. 5.3.3, by using the improved Sanov theorem and some norm inequalities, we derived upper bounds for a maximum-likelihood estimator. We also explained the importance of an improvement of the bounds. The improvement is a topic for future research.

Appendix

A Linear Algebra and Convex Analysis

In this section, we give mathematical supplements. In Sect. A.1, we explain terminologies in vector and matrix theory and known inequalities between different norms. In Sect. A.2, we explain a known property of projections in convex analysis.

A.1 Vector and Matrix

A.1.1 Class of Matrices

Let $M_{m,n}$ denote the set of all complex matrices with m columns and n rows. When $m = n$, the elements of $M_{m,m}$ is called square matrices.

1. Square matrix
   Suppose that a matrix A is square. Let $A^\dagger$ denote the Hermitian conjugate of A. When the matrix A satisfies $A^\dagger A = AA^\dagger$, it is called normal. The normality is the necessary and sufficient condition of the diagonalizability. When A is diagonalizable, it can be represented as

$$A = \sum_{i=1}^{m} a_i v_i v_i^\dagger, \tag{5.150}$$

where $\boldsymbol{v}_i$ are normalized vectors in $\mathbb{C}^m$ and orthogonal to each other. These vectors are called the eigenvectors of A, and the complex values a_i are called the eigenvalues of A. When a square matrix A satisfies $A = A^\dagger$, it is called Hermitian. Hermitian matrices are normal, and the eigenvalues are real. When all eigenvalues of A is positive, it is called a positive matrix. When the eigenvalues are nonnegative, it is called a positive semidefinite matrix.
Let f denote a function from $\mathbb{C}$ to $\mathbb{C}$. Let A denote a normal matrix and $A = \sum_{i=1}^{m} a_i \boldsymbol{v}_i \boldsymbol{v}_i^\dagger$ be the diagonalized form. We define the action of f on A by

$$f(A) := \sum_{i=1}^{m} f(a_i) \boldsymbol{v}_i \boldsymbol{v}_i^\dagger. \tag{5.151}$$

We use the following notation.

$$|A| := \sum_{i=1}^{m} |a_i| \boldsymbol{v}_i \boldsymbol{v}_i^\dagger, \tag{5.152}$$

$$\sqrt{A} :=_{i=1}^{m} \sqrt{a_i} \boldsymbol{v}_i \boldsymbol{v}_i^\dagger. \tag{5.153}$$

When all eigenvalues of A are not zero, we define the inverse matrix of A by

$$A^{-1} := \sum_{i=1}^{m} \frac{1}{a_i} \boldsymbol{v}_i \boldsymbol{v}_i^\dagger. \tag{5.154}$$

When the inverse matrix of A exists, A is called invertible or regular. The inverse matrix satisfies $AA^{-1} = A^{-1}A = I_m$, where I_m is the identity matrix on $\mathbb{C}^m$.

When A has zero eigenvalues, we cannot define the inverse matrix and A is called irregular. In this case, we define a matrix A^- by

$$A^- := \sum_{i; a_i \neq 0} \frac{1}{a_i} \boldsymbol{v}_i \boldsymbol{v}_i^\dagger. \tag{5.155}$$

This A^- is a Moore-Penrose generalized inverse for normal matrices.

2. Non-square matrix
Next, we consider the case of $m \neq n$. For a given non-square matrix $A \in M_{m,n}$, we define the left-inverse matrix $A_{\text{left}}^{-1} \in M_{n,m}$ and right-inverse matrix $A_{\text{right}}^{-1} \in M_{n,m}$ by

$$A_{\text{left}}^{-1} A = I_n, \tag{5.156}$$

$$A A_{\text{right}}^{-1} = I_m. \tag{5.157}$$

We define the rank of A by the number of independent row vectors of A and use the notation $\text{rank}(A)$. When $\text{rank}(A) = \min\{m, n\}$ holds, A is called full-rank, and otherwise it is called rank-deficient. Suppose that A is full-rank. When $m > n$, the left-inverse matrix A_{left}^{-1} exists, and when $m < n$, the right-inverse matrix A_{right}^{-1} exists. When A is rank-deficient, there are no left- and right-inverse matrices of A.

A.1.2 Inhomogeneous Equation

Let A be a real (m, n)-matrix and $\boldsymbol{v} \in \mathbb{R}^n$, $\boldsymbol{w} \in \mathbb{R}^m$. When $\boldsymbol{w}$ is not $\boldsymbol{0}$, an equation

$$A\boldsymbol{v} = \boldsymbol{w} \tag{5.158}$$

is called an inhomogeneous equation. e consider an inverse problem such that for given A and $\boldsymbol{w}$, we find $\boldsymbol{v}$ satisfying Eq. (5.158). We define the image of A by

$$\mathrm{Im}(A) := \{A\boldsymbol{v}' | \boldsymbol{v}' \in \mathbb{R}^n\} \subset \mathbb{R}^m. \tag{5.159}$$

When $\boldsymbol{w} \in \mathrm{Im}(A)$, the solutions of Eq. (5.158) exist, and otherwise they do not exist. We define an augmented matrix of A with $\boldsymbol{w}$ by

$$[A|\boldsymbol{w}] := \left(\begin{array}{ccc|c} A_{1,1} & \cdots & A_{1,n} & w_1 \\ \vdots & \ddots & \vdots & \vdots \\ A_{m,1} & \cdots & A_{m,n} & w_m \end{array}\right). \tag{5.160}$$

Then following theorem holds.

Theorem 5.12 The following statements are equivalent.

(i) For a given A and $\boldsymbol{w}$, Eq. (5.158) has solutions.
(ii) $\boldsymbol{w} \in \mathrm{Im}(A)$.
(iii) The ranks of A and the augmented matrix are same, i.e.,

$$\mathrm{rank}(A) = \mathrm{rank}([A|\boldsymbol{w}]). \tag{5.161}$$

When the solutions exist, the solutions have $\big(n - \mathrm{rank}(A)\big)$ degrees of freedom.

We summarize the contents of this subsection. Suppose that $A \in M_{m,n}$ and $\boldsymbol{w} \in \mathbb{R}^m$ are given.

- If and only if the statements (ii) or (iii) are not satisfied, Eq. (5.158) has no solutions.
- If and only if he statements (ii) or (iii) are satisfied, Eq. (5.158) has solutions.
- If Eq. (5.158) has solutions, they have $\big(n - \mathrm{rank}(A)\big)$ degrees of freedom.
- Suppose that Eq. (5.158) has a solution and $n > \mathrm{rank}(A)$ holds. Then the solution is not unique.
- Suppose that Eq. (5.158) has a solution and $n = \mathrm{rank}(A)$ holds. Then the solution is unique.

– When $n = m$, A is full-rank and invertible, and the solution is given as $A^{-1}\boldsymbol{w}$.
– When $m > n$, A is full-rank and the left-inverse matrix exists. The solution is given as $A_{\text{left}}^{-1}\boldsymbol{w}$.

A.1.3 Norms and Loss Functions

We obey the definitions of norms in [50].

- Vector norms
  A function $\|\cdot\| : \mathbb{C}^k \to \mathbb{R}$ is a vector norm if for all $\boldsymbol{v},\ \boldsymbol{w} \in \mathbb{C}^k$,

(i) (Non-negativity) $\|\boldsymbol{v}\| \geq 0$.
(ii) (Positivity) $\|\boldsymbol{v}\| = 0$ if and only if $\boldsymbol{x} = \boldsymbol{0}$.
(iii) (Homogeneity) $\|c\boldsymbol{v}\| = |c| \cdot \|\boldsymbol{v}\|$ for all scalars $c \in \mathbb{C}$.
(iv) (Triangle inequality) $\|\boldsymbol{v} + \boldsymbol{w}\| \leq \|\boldsymbol{v}\| + \|\boldsymbol{w}\|$.

We introduce three representative vector norms:

1. ℓ_1-norm (the sum norm) $\|\boldsymbol{v}\|_1 := \sum_{i=1}^{k} |v_i|$.
2. ℓ_2-norm (the Euclidean norm) $\|\boldsymbol{v}\|_2 := \sqrt{\sum_{i=1}^{k} |v_i|^2}$.
3. ℓ_∞-norm (the max norm) $\|\boldsymbol{v}\|_\infty := \max_{i=1,\ldots,k} |v_i|$.

These vector norms satisfy the following inequalities:

$$\|\boldsymbol{v}\|_2 \leq \|\boldsymbol{v}\|_1 \leq \sqrt{k}\|\boldsymbol{v}\|_2, \tag{5.162}$$

$$\|\boldsymbol{v}\|_\infty \leq \|\boldsymbol{v}\|_1 \leq k\|\boldsymbol{v}\|_\infty, \tag{5.163}$$

$$\|\boldsymbol{v}\|_\infty \leq \|\boldsymbol{v}\|_2 \leq \sqrt{k}\|\boldsymbol{v}\|_\infty. \tag{5.164}$$

For any probability distributions $\boldsymbol{p}$ and $\boldsymbol{q}$, ℓ_1-norm and the Kullback-Leibler divergence satisfy the following inequality [51]

$$(\|\boldsymbol{p} - \boldsymbol{q}\|_1)^2 \leq K(\boldsymbol{p}\|\boldsymbol{q}). \tag{5.165}$$

Equation (5.165) is called Pinsker's inequality.

- Matrix norms
  A function $|||\cdot||| : M_{k,k} \to \mathbb{R}$ is a matrix norm if for all $A,\ B \in M_{k,k}$,

(i) (Non-negativity) $|||A||| \geq 0$.
(ii) (Positivity) $|||A||| = 0$ if and only if $A = O$.
(iii) (Homogeneity) $|||cA||| = |c| \cdot |||A|||$ for all scalars $c \in \mathbb{C}$.
(iv) (Triangle inequality) $|||A + B||| \leq |||A||| + |||B|||$.
(v) (Submultiplicativity) $|||AB||| \leq |||A||| \cdot |||B|||$.

We introduce two representative matrix norms:

1. Trace norm $|||A|||_{\text{tr}} := \text{tr}[|A|]$.
2. Hilbert-Schmidt norm (the Frobenius norm) $|||A|||_{\text{HS}} := \text{tr}[A^\dagger A]^{1/2}$.

These matrix norms satisfy the following inequalities:

$$|||A|||_{\mathrm{HS}} \leq |||A|||_{\mathrm{tr}} \leq \sqrt{\mathrm{rank}(A)}|||A|||_{\mathrm{HS}}. \tag{5.166}$$

The Hilbert-Schmidt distance between two density matrices are defined by

$$\Delta^{\mathrm{HS}}(\hat{\rho}, \hat{\rho}') := \frac{1}{\sqrt{2}}|||\hat{\rho} - \hat{\rho}'|||_{\mathrm{HS}}. \tag{5.167}$$

The normalization factor makes the maximal value one. In the generalized Bloch parametrization, we have

$$\Delta^{\mathrm{HS}}(\hat{\rho}(\boldsymbol{s}), \hat{\rho}(\boldsymbol{s}')) = \frac{1}{2}\|\boldsymbol{s} - \boldsymbol{s}'\|_2. \tag{5.168}$$

- Fidelity, infidelity, Bures distance
  In quantum information science, one of the most popular evaluation functions is fidelity. For two density matrices $\hat{\rho}$ and $\hat{\rho}'$, the fidelity is defined by

$$f(\hat{\rho}, \hat{\rho}') := \mathrm{Tr}\left[\sqrt{\sqrt{\hat{\rho}}\hat{\rho}'\sqrt{\hat{\rho}}}\right]^2. \tag{5.169}$$

Fidelity satisfies $f(\hat{\rho}, \hat{\rho}) = 1$ and it is not a loss function. The square-root $\sqrt{f(\hat{\rho}, \hat{\rho}')}$ is called the root fidelity. The infidelity is defined by

$$\begin{aligned}\Delta^{\mathrm{IF}}(\hat{\rho}, \hat{\rho}') &:= 1 - \mathrm{Tr}\left[\sqrt{\sqrt{\hat{\rho}}\hat{\rho}'\sqrt{\hat{\rho}}}\right]^2 && (5.170)\\ &= 1 - f(\hat{\rho}, \hat{\rho}'). && (5.171)\end{aligned}$$

Infidelity is a loss function but it is not a distance. The Bures distance is defined by

$$\begin{aligned}\Delta^{\mathrm{B}}(\hat{\rho}, \hat{\rho}') &:= \sqrt{1 - \mathrm{Tr}\left[\sqrt{\sqrt{\hat{\rho}}\hat{\rho}'\sqrt{\hat{\rho}}}\right]} && (5.172)\\ &= \sqrt{1 - \sqrt{f(\hat{\rho}, \hat{\rho}')}}. && (5.173)\end{aligned}$$

For the trace distance, Bures distance, and infidelity, the following inequalities hold [52]:

$$\begin{aligned}\Delta^{\mathrm{B}}(\hat{\rho}, \hat{\rho}') \leq \sqrt{\Delta^{\mathrm{IF}}(\hat{\rho}, \hat{\rho}')} \leq \sqrt{2}\Delta^{\mathrm{B}}(\hat{\rho}, \hat{\rho}'), && (5.174)\\ \Delta^{\mathrm{B}}(\hat{\rho}, \hat{\rho}')^2 \leq \Delta^{\mathrm{T}}(\hat{\rho}, \hat{\rho}') \leq \sqrt{\Delta^{\mathrm{IF}}(\hat{\rho}, \hat{\rho}')}. && (5.175)\end{aligned}$$

A.2 Projection in Convex Analysis

A subset $R \subseteq \mathbb{R}^k$ is called convex if for all $\boldsymbol{r},\ \boldsymbol{r}' \in R$, $p\boldsymbol{r} + (1-p)\boldsymbol{r}'$ is included in S, where $p \in [0, 1]$. Let R be a non-empty, closed, and convex set in $\mathbb{R}^k$ and $\boldsymbol{t}$ be a vector in $\boldsymbol{R}^k$. The vector $\boldsymbol{t}$ is not necessary included in R. We define the projection of $\boldsymbol{t}$ onto R by

$$\boldsymbol{P}_R(\boldsymbol{t}) := \operatorname*{argmin}_{\boldsymbol{r}\in R} \|\boldsymbol{t} - \boldsymbol{r}\|_2. \tag{5.176}$$

Obviously, $\boldsymbol{P}_R(\boldsymbol{t}) = \boldsymbol{t}$ holds for any $\boldsymbol{t} \in R$.

Theorem 5.13 (Non-expandability of projections [53]**)**
Suppose that R is a non-empty, closed, and convex set in $\mathbb{R}^k$. Then for any $\boldsymbol{t}, \boldsymbol{t}' \in \mathbb{R}^k$,

$$\|\boldsymbol{P}_R(\boldsymbol{t}) - \boldsymbol{P}_R(\boldsymbol{t}')\|_2 \le \|\boldsymbol{t} - \boldsymbol{t}'\|_2 \tag{5.177}$$

holds.

Theorem 5.13 indicates that any projections do not expand the distance between any two vectors in the Euclidean space. For $\boldsymbol{t}' \in R$, we have

$$\|\boldsymbol{P}_R(\boldsymbol{t}) - \boldsymbol{t}'\|_2 \le \|\boldsymbol{t} - \boldsymbol{t}'\|_2. \tag{5.178}$$

B Supplement for Sect. 5.3.2

In this section, we give supplements for Sect. 5.3.2. In Sect. B.1, we derive the upper bound on error probability for k-qubit state tomography using Pauli measurements with detection losses. In Sects. B.2 and B.3, we explain a way of evaluating the effect of systematic and numerical errors on the error threshold and upper bound. In Sect. B.4, we derive the upper bound on error probability for the constrained least squares estimator, which was introduced in Sect. 5.3.2, and compare the performance to that of the ℓ_2-eNM estimator.

B.1 Proof of Eq. (5.130)

Suppose that we prepare N identical copies of $\hat{\rho} \in \mathcal{S}((\mathbb{C}^2)^{\otimes k})$ and make the three Pauli measurements with detection efficiency η on each qubit. The POVMs describing the ideal Pauli measurements on each qubit, $\boldsymbol{\Pi}^{(i)} = \{\hat{\Pi}^{(i)}_{+1}, \hat{\Pi}^{(i)}_{-1}\}$, are given as

$$\hat{\Pi}^{(i)}_{\pm 1} := \frac{1}{2}\left(\hat{\mathbb{1}} \pm \boldsymbol{e}_i \cdot \boldsymbol{\sigma}\right), \tag{5.179}$$

where $i = 1, 2, 3,$

$$e_1 := \begin{pmatrix} 1 \\ 0 \\ 0 \end{pmatrix}, \; e_2 := \begin{pmatrix} 0 \\ 1 \\ 0 \end{pmatrix}, \; e_3 := \begin{pmatrix} 0 \\ 0 \\ 1 \end{pmatrix}, \tag{5.180}$$

and

$$\hat{\sigma}_1 := \begin{pmatrix} 0 & 1 \\ 1 & 0 \end{pmatrix}, \; \hat{\sigma}_2 := \begin{pmatrix} 0 & -i \\ i & 0 \end{pmatrix}, \; \hat{\sigma}_3 := \begin{pmatrix} 1 & 0 \\ 0 & -1 \end{pmatrix}. \tag{5.181}$$

When the measurements have detention loss, the corresponding POVMs, $\boldsymbol{\Pi}^{\eta,(i)} = \{\hat{\Pi}^{\eta,(i)}_{+1}, \hat{\Pi}^{\eta,(i)}_{-1}, \hat{\Pi}^{\eta,(i)}_{0}\}$, are given as

$$\hat{\Pi}^{\eta,(i)}_{\pm 1} := \frac{\eta}{2}\left(\hat{\mathbb{1}} \pm \boldsymbol{e}_i \cdot \boldsymbol{\sigma}\right) \tag{5.182}$$

$$\hat{\Pi}^{\eta,(i)}_{0} := (1-\eta)\hat{\mathbb{1}}, \tag{5.183}$$

where η is the detection efficiency and takes the value from 0 to 1. The outcome "0" means no detection at the measurement trial. When we perform the imperfect Pauli measurements on each qubit, the POVM on k-qubit is given as

$$\boldsymbol{\Pi}^{\eta,(\boldsymbol{i})} := \otimes_{q=1}^{k} \boldsymbol{\Pi}^{\eta,(i_q)}, \tag{5.184}$$

where $\boldsymbol{i} = \{i_q\}_{q=1}^{k}$ and $i_q = 1, 2, 3$. The label of the different POVMs, $\boldsymbol{i}$, corresponds to j in this chapter. Suppose that we perform each measurement described by $\boldsymbol{\Pi}^{\eta,(\boldsymbol{i})}$ equally $n := N/3^k$ times.

Let us choose $\boldsymbol{\lambda}$ to be the set of tensor products of Pauli and identity matrices with the normalization factor $1/\sqrt{2^{k-1}}$, *i.e.*,

$$\hat{\lambda}_{\boldsymbol{\beta}} := \frac{1}{\sqrt{2^{k-1}}} \otimes_{q=1}^{k} \hat{\sigma}_{\beta_q}, \tag{5.185}$$

where $\boldsymbol{\beta} := \{\beta_q\}_{q=1}^{k}$ and $\beta_q = 0, 1, 2, 3$. We eliminate from $\boldsymbol{\beta}$ the case that all β_q are 0. The label of the matrices, $\boldsymbol{\beta}$, corresponds to α in this chapter. Using this $\boldsymbol{\lambda}$, any density matrices are represented as

$$\hat{\rho} = \frac{1}{2^k}\hat{\mathbb{1}} + \frac{1}{2}\boldsymbol{\lambda} \cdot \boldsymbol{s}, \tag{5.186}$$

where

$$s_{\boldsymbol{\beta}} = \mathrm{Tr}\left[\hat{\rho}\hat{\lambda}_{\boldsymbol{\beta}}\right] \tag{5.187}$$
$$= \frac{1}{\sqrt{2^{k-1}}}\mathrm{Tr}\left[\hat{\rho}\left(\otimes_{q=1}^{k}\hat{\sigma}_{\beta_q}\right)\right]. \tag{5.188}$$

Equation (5.188) indicates that the parameter $s_{\boldsymbol{\beta}}$ is the expectation of a tensor product of ideal Pauli and identity matrices.

In k-qubit state tomography with $k \geq 2$, we need to be careful about the treatment of multiple uses of same data. For example, in order to estimate the expectation of $\hat{\sigma}_1 \otimes \hat{\mathbb{1}}$ in 2-qubit case, we use the data of three types of measurements; $\hat{\sigma}_1 \otimes \hat{\sigma}_1$, $\hat{\sigma}_1 \otimes \hat{\sigma}_2$, and $\hat{\sigma}_1 \otimes \hat{\sigma}_3$. Therefore the estimation of each parameter can de dependent even for $\hat{\rho}^{\mathrm{LLS}}$.

We try to estimate these parameters from a data set of the imperfect Pauli measurements $\breve{\boldsymbol{\Pi}} := \{\boldsymbol{\Pi}^{\eta,(i)}\}_{\boldsymbol{i}}$. In order to calculate c_α, we need to derive a matrix B satisfying

$$\boldsymbol{s} = B(\boldsymbol{p} - \boldsymbol{a}_0). \tag{5.189}$$

This matrix B corresponds to $\Lambda_{\mathrm{left}}^{-1}$ in this chapter. Let l denote the number of $\hat{\mathbb{1}}$ appearing in $\hat{\lambda}_{\boldsymbol{\beta}}$. The number of $\hat{\lambda}_{\boldsymbol{\beta}}$ including l identities is $3^{k-l} \times \frac{k!}{l!(k-l)!}$. $\hat{\lambda}_{\boldsymbol{\beta}} = \hat{\mathbb{1}}^{\otimes l} \otimes \left(\otimes_{q=l+1}^{k}\hat{\sigma}_{i_q}\right)/\sqrt{2^{k-1}}$ is an example of such $\hat{\lambda}_{\boldsymbol{\beta}}$. In this case, Eq. (5.188) is rewritten by the probability distributions of the imperfect Pauli measurement as

$$s_{\boldsymbol{\beta}} = \frac{1}{\sqrt{2^{k-1}}}\sum_{\substack{m_{l+1},\ldots,m_k;\\ m_q=\pm 1}}\left(\prod_{q=l+1}^{k} m_q\right) p(m_{l+1},\ldots,m_k|\boldsymbol{I}^{\otimes l}$$
$$\otimes(\otimes_{q=l+1}^{k}\boldsymbol{\Pi}^{(i_q)}),\hat{\rho}) \tag{5.190}$$
$$= \frac{1}{\sqrt{2^{k-1}}}\sum_{\substack{m_{l+1},\ldots,m_k;\\ m_q=\pm 1}}\left(\prod_{q=l+1}^{k} m_q\right)\frac{1}{\eta^{k-l}} p(m_{l+1},\ldots,m_k|\boldsymbol{I}^{\otimes l}$$
$$\otimes(\otimes_{q=l+1}^{k}\boldsymbol{\Pi}^{\eta,(i_q)}),\hat{\rho}) \tag{5.191}$$
$$= \frac{1}{\sqrt{2^{k-1}}}\sum_{\substack{i_1,\ldots,i_l;\\ i_q=1,2,3}}\sum_{\substack{m_1,\ldots,m_l;\\ m_q=\pm 1,0}}\sum_{\substack{m_{l+1},\ldots,m_k;\\ m_q=\pm 1}}\left(\prod_{q=l+1}^{k} m_q\right)$$
$$\frac{1}{\eta^{k-l}}\frac{1}{3^l} p(m_1,\ldots,m_k|\boldsymbol{\Pi}^{\eta,(\boldsymbol{i})},\hat{\rho}). \tag{5.192}$$

Therefore we have

$$B_{\boldsymbol{\beta},(\boldsymbol{i},\boldsymbol{m})} = \pm\frac{1}{\sqrt{2^{k-1}}}\frac{1}{\eta^{k-l}}\frac{1}{3^l}, \tag{5.193}$$

if $i_q = 1, 2, 3$ and $m_q = \pm 1, 0$ for $q = 1, \ldots, l$ and $i_q = \beta_q$ and $m_q = \pm 1$ for $q = l+1, \ldots, k$. Otherwise $B_{\boldsymbol{\beta},(\boldsymbol{i},\boldsymbol{m})} = 0$. Then for each $\boldsymbol{\beta}$ and $\boldsymbol{i}$,

$$\max_{\boldsymbol{m}} B_{\boldsymbol{\beta},(\boldsymbol{i},\boldsymbol{m})} - \min_{\boldsymbol{m}} B_{\boldsymbol{\beta},(\boldsymbol{i},\boldsymbol{m})} = \begin{cases} \frac{2}{\sqrt{2^{k-1}}\eta^{k-l}3^{l}} & \text{if } i_q = \beta_q \text{ for } q = l+1, \ldots, k \\ 0 & \text{otherwise} \end{cases} \tag{5.194}$$

holds, and we obtain

$$c_{\boldsymbol{\beta}} = \sum_{\boldsymbol{i}} 3^k \left\{ \max_{\boldsymbol{m}} B_{\boldsymbol{\beta},(\boldsymbol{i},\boldsymbol{m})} - \min_{\boldsymbol{m}} B_{\boldsymbol{\beta},(\boldsymbol{i},\boldsymbol{m})} \right\}^2 \tag{5.195}$$

$$= \sum_{\substack{i_1,\ldots,i_l; \\ i_q = 1,2,3}} 3^k \left\{ \frac{2}{\sqrt{2^{k-1}}\eta^{k-l}3^{l}} \right\}^2 \tag{5.196}$$

$$= \frac{3^{k-l}}{2^{k-3}\eta^{2(k-l)}}. \tag{5.197}$$

From the above discussion, we can see that $c_{\boldsymbol{\beta}}$ takes same value for different $\hat{\lambda}_{\boldsymbol{\beta}}$ with the same l. The upper bound of error probability is calculated as

$$\begin{aligned} \mathrm{P}_u^{\Delta}(\delta, N, \breve{\boldsymbol{\Pi}}, \hat{\rho}^{\ell_2\text{-eNM}}) &= 2 \sum_{\boldsymbol{\beta}} \exp\left[-\frac{b}{c_{\boldsymbol{\beta}}} \delta^2 N \right] \\ &= 2 \sum_{l=0}^{k-1} 3^{k-l} \binom{k}{l} \exp\left[-b \frac{2^{k-3}\eta^{2(k-l)}}{3^{k-l}} \delta^2 N \right], \end{aligned} \tag{5.198}$$

where

$$b := \begin{cases} 8/(d^2-1) & \text{if } \Delta = \Delta^{\mathrm{HS}} \\ 16/d(d^2-1) & \text{if } \Delta = \Delta^{\mathrm{T}} \\ 4/d(d^2-1) & \text{if } \Delta = \Delta^{\mathrm{IF}} \end{cases}. \tag{5.199}$$

When we choose the trace distance as the loss function, we have

$$b = \frac{16}{d(d^2-1)} = \frac{1}{2^{k-4} \cdot (2^{2k}-1)}, \tag{5.200}$$

and

$$\mathrm{P}_u^{\mathrm{T}}(\delta, N, \breve{\boldsymbol{\Pi}}, \hat{\rho}^{\ell_2\text{-eNM}}) = 2 \sum_{l=0}^{k-1} 3^{k-l} \binom{k}{l} \exp\left[-\frac{2}{2^{2k}-1} \frac{\eta^{2(k-l)}}{3^{k-l}} \delta^2 N \right]. \tag{5.201}$$

In one-qubit ($k = 1$) and two-qubit ($k = 2$) cases, we have

$$\mathrm{P}_u^{\mathrm{T}}(k=1) = 6\exp\left[-\frac{2}{9}\eta^2\delta^2 N\right], \tag{5.202}$$

$$\mathrm{P}_u^{\mathrm{T}}(k=2) = 18\exp\left[-\frac{2}{135}\eta^4\delta^2 N\right] + 12\exp\left[-\frac{2}{45}\eta^2\delta^2 N\right]. \tag{5.203}$$

As in the above discussion, when the directions of each Pauli measurement are perfectly orthogonal, it is easy to derive $c_{\boldsymbol{\beta}}$. When the directions are not orthogonal, we need to calculate $\Lambda_{\mathrm{left}}^{-1} = (\Lambda^T \Lambda)^{-1}\Lambda^T$. Then, it becomes more difficult to analyze $c_{\boldsymbol{\beta}}$, and we would need to calculate them numerically.

B.2 Effect of Systematic Errors

Theorems 5.9 is valid for any informationally complete POVMs and is applicable for cases in which a systematic error exists. However, we must know exactly the mathematical representation of the systematic error in order to strictly verify a value of the confidence level. This assumption can be unrealistic in some experiments. In this section, we will weaken the assumption to a more realistic condition and give a formula of P_u^{Δ} in such a case.

Let $\breve{\boldsymbol{\Pi}}$ denote a set of POVMs exactly describing the measurement used, and let $\breve{\boldsymbol{\Pi}}'(\neq \breve{\boldsymbol{\Pi}})$ denote a set of POVMs that we mistake as the correct set of POVMs. We assume that $\breve{\boldsymbol{\Pi}}$ and $\breve{\boldsymbol{\Pi}}'$ are both informationally complete. Suppose that we do not know $\breve{\boldsymbol{\Pi}}$, but we know that $\breve{\boldsymbol{\Pi}}$ is in a known set $\mathscr{M}$. For example, consider the case where an experimentalist wants to perform a projective easurement of $\hat{\sigma}_1$. If they can guarantee that their actual measurement is prepared within 0.5 degrees from the x-axis, and if their detection efficiency is 0.9, then $\mathscr{M}$ is the set of all POVMs whose measurement direction and detection efficiency are within 0.5 degrees of the x-axis and 0.9, respectively.

For given relative frequencies $\boldsymbol{\nu}_N$, the correct and mistaken eL estimates are

$$\boldsymbol{s}_N^{\mathrm{eL}} = \Lambda_{\mathrm{left}}^{-1}(\breve{\boldsymbol{\Pi}})\left\{\boldsymbol{\nu}_N - \boldsymbol{a}_0\right\}, \tag{5.204}$$

$$\boldsymbol{s}_N^{\mathrm{eL}\prime} = \Lambda_{\mathrm{left}}^{-1}(\breve{\boldsymbol{\Pi}}')\left\{\boldsymbol{\nu}_N - \boldsymbol{a}_0'\right\}. \tag{5.205}$$

Then the actual and mistaken ℓ_2-eNM estimates are

$$\boldsymbol{s}_N^{\ell_2\text{-eNM}} = \operatorname*{argmin}_{\boldsymbol{s}' \in B_d} \|\boldsymbol{s}' - \boldsymbol{s}_N^{\mathrm{eL}}\|_2, \tag{5.206}$$

$$\boldsymbol{s}_N^{\ell_2\text{-eNM}\prime} = \operatorname*{argmin}_{\boldsymbol{s}' \in B_d} \|\boldsymbol{s}' - \boldsymbol{s}_N^{\mathrm{eL}\prime}\|_2. \tag{5.207}$$

Let $\hat{\rho}_N^{\ell_2\text{-eNM}}$ and $\hat{\rho}_N^{\ell_2\text{-eNM}\prime}$ denote the corresponding density matrix estimates. Let us define the size of the systematic error as

$$\xi := \max_{\breve{\boldsymbol{\Pi}} \in \mathscr{M}} \Delta(\hat{\rho}_N^{\ell_2\text{-eNM}\prime}, \hat{\rho}_N^{\text{eNM}}). \tag{5.208}$$

This is a function of Δ, $\boldsymbol{\nu}_N$, $\breve{\boldsymbol{\Pi}}'$, and $\mathscr{M}$. Then for any $\hat{\rho} \in \mathscr{S}(\mathscr{H})$ and $\breve{\boldsymbol{\Pi}} \in \mathscr{M}$,

$$\begin{aligned}\Delta(\hat{\rho}_*, \hat{\rho}) &\leq \Delta(\hat{\rho}_*, \hat{\rho}_N^{\ell_2\text{-eNM}\prime}) + \Delta(\hat{\rho}_N^{\ell_2\text{-eNM}\prime}, \hat{\rho}_N^{\ell_2\text{-eNM}}) + \Delta(\hat{\rho}_N^{\ell_2\text{-eNM}}, \hat{\rho}) \\ &\leq \Delta(\hat{\rho}_*, \hat{\rho}_N^{\ell_2\text{-eNM}\prime}) + \xi + \delta \end{aligned} \tag{5.209}$$

holds with probability at least

$$1 - \min_{\breve{\boldsymbol{\Pi}} \in \mathscr{M}} \mathrm{P}_u^{\Delta} = 1 - 2 \max_{\breve{\boldsymbol{\Pi}} \in \mathscr{M}} \sum_{\alpha=1}^{d^2-1} \exp\left[-\frac{b}{c_\alpha} \delta^2 N \right]. \tag{5.210}$$

Using Eqs. (5.209) and (5.210), we can evaluate the precision of state preparation, $\Delta(\hat{\rho}_*, \hat{\rho})$, without knowing the true state $\hat{\rho}$ and true sets of POVMs $\breve{\boldsymbol{\Pi}}$.

B.3 Effect of Numerical Errors

In this section, we analyze the effect of numerical errors and explain a method for evaluating the precision of the state preparation in the cases that numerical errors exits.

The ℓ_2-eNM estimator $\hat{\rho}^{\ell_2\text{-eNM}}$ requires a nonlinear minimization, which requires the use of a numerical algorithm. Suppose that we choose an algorithm for the minimization and obtain a result $\hat{\sigma}_N^{\ell_2\text{-eNM}}$ for a given data set. In practice, there exists a numerical error on the result, and $\hat{\sigma}_N^{\ell_2\text{-eNM}}$ differs from the exact solution $\hat{\rho}_N^{\ell_2\text{-eNM}}$. We cannot obtain the exact solution, but we can guarantee the accuracy of the numerical result with accuracy-guaranteed algorithms [54]. Suppose that we use an algorithm for which $\Delta(\hat{\sigma}_N^{\ell_2\text{-eNM}}, \hat{\rho}_N^{\ell_2\text{-eNM}}) \leq \zeta$ is guaranteed. Then

$$\begin{aligned}\Delta(\hat{\rho}_*, \hat{\rho}) &\leq \Delta(\hat{\rho}_*, \hat{\sigma}_N^{\ell_2\text{-eNM}}) + \Delta(\hat{\sigma}_N^{\ell_2\text{-eNM}}, \hat{\rho}_N^{\ell_2\text{-eNM}}) + \Delta(\hat{\rho}_N^{\ell_2\text{-eNM}}, \hat{\rho}) \\ &\leq \Delta(\hat{\rho}_*, \hat{\sigma}_N^{\ell_2\text{-eNM}}) + \zeta + \delta \end{aligned} \tag{5.211}$$

holds with probability at least $1 - \mathrm{P}_u^{\Delta}$. The error threshold is changed from δ to $\zeta + \delta$.

Usually systematic and numerical errors both exists. In such a case, by combining Eqs. (5.209) and (5.211), we can prove that the inequality

$$\Delta(\hat{\rho}_*, \hat{\rho}) \leq \Delta(\hat{\rho}_*, \hat{\sigma}_N^{\ell_2\text{-eNM}}) + \zeta + \xi + \delta \tag{5.212}$$

holds with probability in Eq. (5.210), where ζ is a numerical error threshold for $\breve{\boldsymbol{\Pi}}'$. Therefore Theorem 5.9 with a modification can apply for the cases that systematic and numerical errors exist.

B.4 Error Probability for Constrained Least Squares Estimator

From Eq. (5.56), the probability distribution of $\hat{\rho}_N^{\mathrm{eL}}$ is the projection of $\boldsymbol{\nu}_N$ on the probability space of trace-one Hermitian matrices ($\{\boldsymbol{p}(\hat{\sigma})|\hat{\sigma}=\hat{\sigma}^\dagger, \mathrm{Tr}[\hat{\sigma}]=1\}$), and we have

$$\|\boldsymbol{p}(\hat{\rho}')-\boldsymbol{\nu}_N\|_2^2 = \|\boldsymbol{p}(\hat{\rho}')-\boldsymbol{p}(\hat{\rho}_N^{\mathrm{eL}})\|_2^2 + \|\boldsymbol{p}(\hat{\rho}_N^{\mathrm{eL}})-\boldsymbol{\nu}_N\|_2^2, \ \forall\hat{\rho}'\in\mathcal{S}(\mathcal{H}). \tag{5.213}$$

Therefore, Eq. (5.131) is rewritten as

$$\hat{\rho}_N^{\mathrm{CLS}} = \operatorname*{argmin}_{\hat{\rho}'\in\mathcal{S}(\mathcal{H})} \|\boldsymbol{p}(\hat{\rho}')-\boldsymbol{p}(\hat{\rho}_N^{\mathrm{eL}})\|_2, \tag{5.214}$$

and $\hat{\rho}_N^{\mathrm{CLS}}$ is the projection of $\hat{\rho}_N^{\mathrm{eL}}$ on $\mathcal{S}(\mathcal{H})$ with respect to the 2-norm on the probability space. We can see from Eqs. (5.68) and (5.214) that $\hat{\rho}^{\ell_2\text{-eNM}}$ and $\hat{\rho}^{\mathrm{CLS}}$ are the projections of $\hat{\rho}_N^{\mathrm{eL}}$ with respect to difference spaces (or different norms).

Using Theorem 5.13, we obtain

$$\|\boldsymbol{p}(\hat{\rho}_N^{\mathrm{CLS}})-\boldsymbol{p}(\hat{\rho})\|_2 \le \|\boldsymbol{p}(\hat{\rho}_N^{\mathrm{eL}})-\boldsymbol{p}(\hat{\rho})\|_2, \ \forall\hat{\rho}\in\mathcal{S}(\mathcal{H}), \tag{5.215}$$

$$\|\Lambda(\boldsymbol{s}_N^{\mathrm{CLS}}-\boldsymbol{s})\|_2 \le \|\Lambda(\boldsymbol{s}_N^{\mathrm{eL}}-\boldsymbol{s})\|_2, \ \forall\boldsymbol{s}\in B_d, \tag{5.216}$$

where $\boldsymbol{s}_N^{\mathrm{CLS}}$ is the Bloch vector corresponding to $\hat{\rho}_N^{\mathrm{CLS}}$. Let us define $\|\Lambda\|_{2,\max}$ and $\|\Lambda\|_{2,\min}$ as

$$\|\Lambda\|_{2,\max} := \max_{\boldsymbol{v}\neq\boldsymbol{0}} \frac{\|\Lambda\boldsymbol{v}\|_2}{\|\boldsymbol{v}\|_2}, \tag{5.217}$$

$$\|\Lambda\|_{2,\min} := \min_{\boldsymbol{v}\neq\boldsymbol{0}} \frac{\|\Lambda\boldsymbol{v}\|_2}{\|\boldsymbol{v}\|_2}. \tag{5.218}$$

When $\breve{\boldsymbol{\Pi}}$ is informationally complete, Λ is full-rank and $\|\Lambda\|_{\min}>0$. We have

$$\|\Lambda\|_{2,\min}\cdot\|\boldsymbol{s}_N^{\mathrm{CLS}}-\boldsymbol{s}\|_2 \le \|\Lambda(\boldsymbol{s}_N^{\mathrm{CLS}}-\boldsymbol{s})\|_2 \tag{5.219}$$

$$\le \|\Lambda(\boldsymbol{s}_N^{\mathrm{eL}}-\boldsymbol{s})\|_2 \tag{5.220}$$

$$\le \|\Lambda\|_{2,\max}\cdot\|\boldsymbol{s}_N^{\mathrm{eL}}-\boldsymbol{s}\|_2. \tag{5.221}$$

We obtain

$$\mathbb{P}\left[\|s_N^{\mathrm{CLS}} - s\|_2 > \delta\right] \le \mathbb{P}\left[\|s_N^{\mathrm{eL}} - s\|_2 > \frac{\|\Lambda\|_{2,\min}}{\|\Lambda\|_{2,\max}}\delta\right]. \tag{5.222}$$

From the same logic in the proof of Theorem 5.14, we obtain the following theorem:

Theorem 5.14 (Error probability, $\hat{\rho}^{\mathrm{CLS}}$, Δ^{HS}, Δ^{T}, Δ^{IF})
When we choose the Hilbert-Schmidt distance, Trace distance, or infidelity as the loss function for the density matrix, we have the following upper bounds on the error probabilities for the constrained least squares estimator.

$$\mathrm{P}^{\mathrm{HS}}_{\delta,N}(\breve{\boldsymbol{\Pi}}, \hat{\rho}^{\mathrm{CLS}}|\hat{\rho}) \le 2\sum_{\alpha=1}^{d^2-1} \exp\left[-\left(\frac{\|\Lambda\|_{2,\min}}{\|\Lambda\|_{2,\max}}\right)^2 \frac{8}{d^2-1}\frac{\delta^2}{c_\alpha}N\right], \tag{5.223}$$

$$\mathrm{P}^{\mathrm{T}}_{\delta,N}(\breve{\boldsymbol{\Pi}}, \hat{\rho}^{\mathrm{CLS}}|\hat{\rho}) \le 2\sum_{\alpha=1}^{d^2-1} \exp\left[-\left(\frac{\|\Lambda\|_{2,\min}}{\|\Lambda\|_{2,\max}}\right)^2 \frac{16}{d(d^2-1)}\frac{\delta^2}{c_\alpha}N\right], \tag{5.224}$$

$$\mathrm{P}^{\mathrm{IF}}_{\delta,N}(\breve{\boldsymbol{\Pi}}, \hat{\rho}^{\mathrm{CLS}}|\hat{\rho}) \le 2\sum_{\alpha=1}^{d^2-1} \exp\left[-\left(\frac{\|\Lambda\|_{2,\min}}{\|\Lambda\|_{2,\max}}\right)^2 \frac{4}{d(d^2-1)}\frac{\delta^2}{c_\alpha}N\right], \tag{5.225}$$

for any true density matrix $\hat{\rho}$.

Compared to Theorem 5.9, there is an additional factor $\left(\frac{\|\Lambda\|_{2,\min}}{\|\Lambda\|_{2,\max}}\right)^2$ (≤ 1) in the rate of exponential decrease in Theorem 5.14. When $\|\Lambda\|_{2,\max} = \|\Lambda\|_{2,\min}$ holds, the upper bounds for $\hat{\rho}^{\mathrm{CLS}}$ coincides with those for $\hat{\rho}^{\ell_2\text{-eNM}}$. Roughly speaking, the condition, $\|\Lambda\|_{2,\max} = \|\Lambda\|_{2,\min}$, implies that we perform measurements extracting information of each Bloch vector element with an equivalent weight. When $\|\Lambda\|_{2,\max} > \|\Lambda\|_{2,\min}$, the upper bounds for $\hat{\rho}^{\mathrm{CLS}}$ is larger than those for $\hat{\rho}^{\ell_2\text{-eNM}}$. This does not mean we can immediately conclude that $\hat{\rho}^{\mathrm{CLS}}$ is less precise than $\hat{\rho}^{\ell_2\text{-eNM}}$ because their upper bounds are probably not optimal. However, we can say that $\hat{\rho}^{\mathrm{CLS}}$ is less precise than $\hat{\rho}^{\ell_2\text{-eNM}}$ insofar as Theorems 5.9 and 5.14 give the only upper bounds known for point estimators in quantum tomography to date. Additionally, the computational cost of $\hat{\rho}^{\ell_2\text{-eNM}}$ can be smaller than that of $\hat{\rho}^{\mathrm{CLS}}$ as explained in Sect. 5.3.2. Therefore, we believe that the ℓ_2-eNM estimator performs better than the CLS estimator and is at present our best choice.

References

1. D.T. Smithey, M. Beck, M.G. Raymer, A. Faridani, Phys. Rev. Lett. **70**, 1244 (1993). doi:10.1103/PhysRevLett.70.1244
2. Z. Hradil, Phys. Rev. A **55**, R1561 (1997). doi:10.1103/PhysRevA.55.R1561
3. K. Banaszek, G.M. D'Ariano, M.G.A. Paris, M.F. Sacchi, Phys. Rev. A **61**, 010304(R) (1999). doi:10.1103/PhysRevA.61.010304

4. J.F. Poyatos, J.I. Cirac, P. Zoller, Phys. Rev. Lett. **78**, 390 (1997). doi:10.1103/PhysRevLett.78.390
5. I.L. Chuang, M.A. Nielsen, J. Mod. Phys. **44**, 2455 (1997). doi:10.1080/09500349708231894
6. V. Buzek, Phys. Rev. A **58**, 1723 (1998). doi:10.1103/PhysRevA.58.1723
7. J. Fiurasek, Z. Hradil, Phys. Rev. A **63**, 020101(R) (2001). doi:10.1103/PhysRevA.63.020101
8. M.F. Sacchi, Phys. Rev. A **63**, 054104 (2001). doi:10.1103/PhysRevA.63.054104
9. A. Luis, L.L. Sanchez-Sato, Phys. Rev. Lett. **83**, 3573 (1999). doi:10.1103/PhysRevLett.83.3573
10. J. Fiurasek, Phys. Rev. A **64**, 024102 (2001). doi:10.1103/PhysRevA.64.024102
11. G.M. D'Ariano, P.L. Presti, Phys. Rev. Lett. **86**, 4195 (2001). doi:10.1103/PhysRevLett.86.4195
12. F. Bloch, Phys. Rev. **70**, 460 (1946). doi:10.1103/PhysRev.70.460
13. E. Bagan, M. Baig, R. Muñoz-Tapia, A. Rodriguez, Phys. Rev. A **69**, 010304(R) (2004). doi:10.1103/PhysRevA.69.010304
14. G. Kimura, Phys. Lett. A **314**, 339 (2003). doi:10.1016/S0375-9601(03)00941--1
15. M.S. Byrd, N. Khaneja, Phys. Rev. A **68**, 062322 (2003). doi:10.1103/PhysRevA.68.062322
16. U. Fano, Rev. Mod. Phys. **29**, 74 (1957). doi:10.1103/RevModPhys.29.74
17. R. Schack, T.A. Brum, C.M. Caves, Phys. Rev. A **64**, 014305 (2001). doi:10.1103/PhysRevA.64.014305
18. C.A. Fuchs, R. Schack, P.F. Scudo, Phys. Rev. A **69**, 062305 (2004). doi:10.1103/PhysRevA.69.062305
19. V. Bužěk, G. Drobny, J. Mod. Opt. **47**, 2823 (2000). doi:10.1080/09500340008232199
20. S.T. Flammia, D. Gross, Y.K. Liu, J. Eisert, New J. Phys. **14**, 095022 (2012). doi:10.1088/1367-2630/14/9/095022
21. R. Blume-Kohout, arXiv:1202.5270 [quant-ph] (2012).
22. M. Christandl, R. Renner, Phys. Rev. Lett. **109**, 120403 (2012). doi:10.1103/PhysRevLett.109.120403
23. A.J. Scott, J. Phys. A: Math. Gen. **39**, 13507 (2006). doi:10.1088/0305-4470/39/43/009
24. H. Zhu, B.G. Englert, Phys. Rev. A **84**, 022327 (2011). doi:10.1103/PhysRevA.84.022327
25. M.D. de Burgh, N.K. Langford, A.C. Doherty, A. Gilchrist, Phys. Rev. A **78**, 052122 (2008). doi:10.1103/PhysRevA.78.052122
26. T. Sugiyama, P.S. Turner, M. Murao, Phys. Rev. A **85**, 052107 (2012). doi:10.1103/PhysRevA.85.052107
27. H. Chernoff, Ann. Math. Stat. **25**, 573 (1954). doi:10.1214/aoms/1177728725
28. S.G. Self, K.Y. Liang, J. Am. Stat. Assoc. **82**, 605 (1987). doi:10.1080/01621459.1987.10478472
29. M. Abramowitz, I.A. Stegun (eds.), *Handbook of Mathematical Functions with Formulas, Graphs, and Mathematical Tables* (Wiley, New York, 1972)
30. J.A. Smolin, J.M. Gambetta, G. Smith, Phys. Rev. Lett. **108**, 070502 (2012). doi:10.1103/PhysRevLett.108.070502
31. S.M. Tan, J. Mod. Opt. **44**, 2233 (1997). doi:10.1080/09500349708231881
32. C.W. Helstrom, *Quantum Detection and Estimation Theory* (Academic, New Tork, 1976)
33. A.S. Holevo, *Probabilistic and Statistical Aspects of Quantum Theory* (North-Holland, New York, 1982)
34. M. Hayashi (ed.), *Asymptotic Theory of Quantum Statistical Inference: Selected Papers* (World Scientific, Singapore, 2005)
35. K. Vogel, H. Risken, Phys. Rev. A **40**, 2847 (1989). doi:10.1103/PhysRevA.40.2847
36. T.M. Buzug, *Computed Tomography: From Photon Statistics to Modern Cone-Beam CT* (Springer, Berlin, 2008)
37. M. Paris, J. Řeháček (eds.), *Quantum State Estimation*. Lecture Notes in Physics (Springer, Berlin, 2004)
38. A. Ling, K.P. Soh, A. L.-Linares, C. Kurtsiefer, Phys. Rev. A **74**, 022309 (2006). doi:10.1103/PhysRevA.74.022309

39. H. Kosaka, T. Inagaki, Y. Rikitake, H. Imamura, Y. Mitsumori, K. Edamatsu, Nature **457**, 702 (2009). doi:10.1038/nature07729
40. M. Steffen, M. Ansmann, R. McDermott, N. Katz, R.C. Bialczak, E. Lucero, Phys. Rev. Lett. **97**, 050502 (2006). doi:10.1103/PhysRevLett.97.050502
41. M. Neeley, M. Ansmann, R.C. Bialczak, M. Hofheinz, N. Katz, E. Lucero, A. O'Connell, H. Wang, A.N. Cleland, J.M. Martinis, Nature Phys. **4**, 523 (2008). doi:10.1038/nphys972
42. M. Hofheinz, H. Wang, M. Ansmann, R.C. Bialczak, E. Lucero, M. Neeley, A.D. O'Connell, D. Sank, J. Wenner, J.M. Martinis, A.N. Cleland, Nature **459**, 546 (2009). doi:10.1038/nature08005
43. D. Leibfried, D.M. Meekhof, B.E. King, C. Monroe, W.M. Itano, D.J. Wineland, Phys. Rev. Lett. **77**, 4281 (1996). doi:10.1103/PhysRevLett.77.4281
44. S. Olmschens, D.N. Matsukevich, P. maunz, D. hayes, L.M. Duan, C. Monroe, Science **323**, 486 (2009). doi:10.1126/science.1167209
45. H. Tanji, S. Ghosh, J. Simon, B. Bloom, V. Vuletic, Phys. Rev. Lett. **103**, 043601 (2009). doi:10.1103/PhysRevLett.103.043601
46. P. Neumann, N. Mizouchi, F. Rempp, P. Hemmer, H. Watanabe, S. Yamasaki, V. Jacques, T. Gaebel, F. Jelezko, J. Wrachtrup, Science **320**, 1326 (2008). doi:10.1126/science.1157233
47. T.J. Dunn, I.A. Walmsley, S. Mukamel, Phys. Rev. Lett. **74**, 884 (1995). doi:10.1103/PhysRevLett.74.884
48. M.A. Nielsen, E. Knill, R. Laflamme, Nature **396**, 52 (1998). doi:10.1038/23891
49. E. Skovsen, H. Stapelfeldt, S. Juhl, K. Molmer, Phys. Rev. Lett. **91**, 090406 (2003). doi:10.1103/PhysRevLett.91.090406
50. R.A. Horn, C.R. Johnson, *Matrix Analysis* (Cambridge University Press, New York, 1985)
51. T.M. Cover, J.A. Thomas, Elements of Information Theory, 2nd edn. Wiley Series in Telecommunications and Signal Processing (Wiley-Interscience, New York, 2006)
52. I. Bengtsson, K. Życzkowski, *Geometry of Quantum States* (Cambridge University Press, Cambridge, 2006)
53. J.M. Borwein, A.S. Lewis, *Convex Analysis and Nonlinear Optimization. CMS Books in Mathematics* (Springer, New York, 2006)
54. N.J. Higham, *Accuracy and Stability of Numerical Algorithms* (SIAM., Philadelphia, 2002).
55. T. Sugiyama, P.S. Turner, M. Murao, Phys. Rev. Lett. **111**, 160406 (2013). doi:10.1103/PhysRevLett.111.160406
56. T. Sugiyama, P.S. Turner, M. Murao, New J. Phys. **14**, 085005 (2012). doi:10.1088/1367-2630/14/8/085005

Chapter 6
Improvement of Estimation Precision by Adaptive Design of Experiments

6.1 Terminology

In this section, we explain basic concepts in adaptive design of experiments for quantum state estimation. Suppose that we have a sequence of N identical copies of an unknown quantum state $\hat{\rho}$ and perform known quantum measurements on each copy. Adaptivity in our sense means that the POVM performed at $(n+1)$-th trial can depend on all the previous n trials' outcomes and POVMs.

6.1.1 *Measurement Update Criterion*

Let $X^{\mathrm{exp}} = \{X_1, \dots, X_N, \cdots\}$ denote a design of experiment in statistical parameter estimation. When the choice of the n-th random variable X_n depends on the previously obtained outcomes $\boldsymbol{x}^{n-1}$, X^{exp} is called an adaptive design of experiment. In quantum state estimation, this corresponds to the case in which during the experiment for state estimation, we change the measurements according to the previously obtained data at each measurement trial.

First let us determine the set of quantum measurements which are available at the n-th trial. We call this set a measurement class at the n-th trial, and let $\mathscr{M}_n$ denote the set of POVMs describing the measurements available at the n-th trial. We choose the n-th POVM, $\boldsymbol{\Pi}_n = \{\hat{\Pi}_{n,x}\}_{x\in\mathscr{X}_n}$ from $\mathscr{M}_n$, where $\mathscr{X}_n$ denotes the set of measurement outcomes for the n-th trial. When the choice is independent of the trial, as in quantum state tomography, we omit the index, using $\mathscr{M}$ for the measurement class and $\mathscr{X}$ for the outcome set. Let $\boldsymbol{x}^n = \{x_1, \dots, x_n\}$ denote the sequence of outcomes obtained up to the n-th trial, where $x_i \in \mathscr{X}_i$. We will denote the pair of measurement performed and outcome obtained by $D_n = (\boldsymbol{\Pi}_n, x_n) \in \mathscr{D}_n := \mathscr{M}_n \times \mathscr{X}_n$, and refer to it as the data for the n-th trial. The sequence of data up to the n-th trial is thus $D^n = \{D_1, \dots, D_n\} \in \mathscr{D}^n := \times_{i=1}^n \mathscr{D}_i$. After the n-th measurement, we choose the next, $(n+1)$-th, POVM $\boldsymbol{\Pi}_{n+1} = \{\Pi_{n+1,x}\}_{x\in\mathscr{X}_{n+1}}$ according to the

T. Sugiyama, *Finite Sample Analysis in Quantum Estimation*, Springer Theses, DOI: 10.1007/978-4-431-54777-8_6, © Springer Japan 2014

previously obtained data. Let u_{n+1} denote the map from the n-th data to the $(n+1)$-th measurement, that is, $u_{n+1} : \mathscr{D}^n \to \mathscr{M}_n$, $\boldsymbol{\Pi}_{n+1} = u_{n+1}(D^n)$. We call u_{n+1} the measurement update criterion for the $(n+1)$-th trial and $u := \{u_1, u_2, \ldots, u_N, \ldots\}$ the measurement update rule. We use the notation $u^N := \{u_1, \ldots, u_N\}$ for the sequence of update criteria used up to the n-th trial. Note that u_1 is a map from $\emptyset$ to $\mathscr{M}_1$ and corresponds to the choice of the first measurement. An adaptive design of experiments is characterized by a measurement update rule u.

6.1.2 Notation

Let $\mathscr{O}$ is the estimation object in quantum estimation. We consider quantum state estimation. The estimation object $\mathscr{O}$ is a set of density matrices on $\mathscr{H}$. If we have some information about the true state, for example, when we know that the true state is pure, the estimation object is a subset of $\mathscr{S}(\mathscr{H})$. In this chapter we consider the case in which we have no information about the true state, that is, $\mathscr{O} = \mathscr{S}(\mathscr{H})$.

Let $\hat{\rho}$ denote the density matrix describing the true state. Suppose that we performed measurements along with an update rule u up to the n-th trial. Then the probability we observe a data sequence $D^N = \{(\boldsymbol{\Pi}_1, x_1), \ldots, (\boldsymbol{\Pi}_N, x_N)\}$ is calculated as

$$p(D^N|u, \hat{\rho}) := \mathrm{Tr}[(\hat{\Pi}_{1,x_1} \otimes \hat{\Pi}_{2,x_2} \otimes \cdots \otimes \hat{\Pi}_{N,x_N})\hat{\rho}^{\otimes N}], \tag{6.1}$$

where

$$\boldsymbol{\Pi}_n := u_n(D^{n-1}),\ n = 1, \ldots, N. \tag{6.2}$$

Let $\hat{\rho}^{\mathrm{est}} = \{\hat{\rho}_1^{\mathrm{est}}, \ldots, \hat{\rho}^{\mathrm{est}}, \ldots\}$ denote an estimator of density matrix. A quintuplet $(\mathscr{O}, N, \mathscr{M}^N, u, \rho^{\mathrm{est}})$ specifies an adaptive estimation scheme. A sketch of the procedure for a generic adaptive quantum estimation scheme is given in Fig. 6.1.

For a given adaptive estimation scheme $(\mathscr{O}, N, \mathscr{M}^N, u, \rho^{\mathrm{est}})$ and a loss function Δ, the expected loss and error probability are defined as

$$\bar{\Delta}_N(u, \hat{\rho}^{\mathrm{est}}|\hat{\rho}) := \sum_{D^N \in \mathscr{D}^N} p(D^N|u, \hat{\rho})\Delta(\hat{\rho}_N^{\mathrm{est}}(D^N), \hat{\rho}), \tag{6.3}$$

$$\mathrm{P}_{\delta,N}(u, \hat{\rho}^{\mathrm{est}}|\hat{\rho}) := \sum_{D^N; \Delta(\hat{\rho}_N^{\mathrm{est}}(D^N), \hat{\rho}) > \delta} p(D^N|u, \hat{\rho}). \tag{6.4}$$

In this chapter, we focus on the analysis of expected losses. We also analyze the average expected loss as

$$\bar{\Delta}_N^{\mathrm{ave}}(u, \hat{\rho}^{\mathrm{est}}) := \int_{\hat{\rho} \in \mathscr{O}} d\mu(\hat{\rho})\bar{\Delta}_N(u, \hat{\rho}^{\mathrm{est}}|\hat{\rho}), \tag{6.5}$$

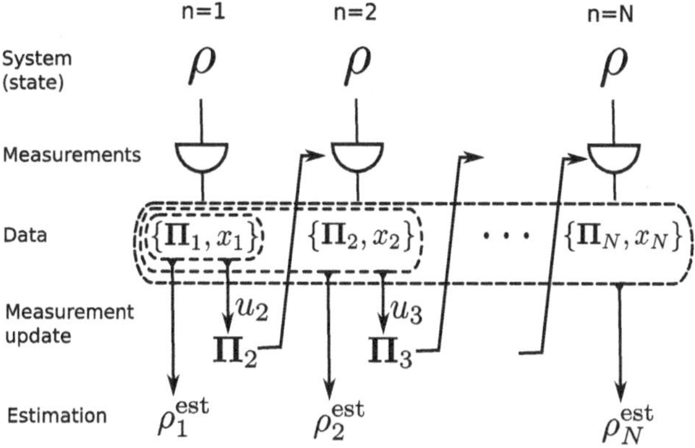

Fig. 6.1 A sketch of a generic procedure for an adaptive quantum estimation scheme. (Reproduced from Ref. [24] with permission)

where μ is a probability measure on $\mathscr{O}$. Let us introduce a parametrization of density matrix $\hat{\rho} = \hat{\rho}(s)$. We use the notation of the pointwise and average expected losses for the parametrization as

$$\bar{\Delta}_N(u, s^{\text{est}}|s), \text{ and } \bar{\Delta}_N^{\text{ave}}(u, s^{\text{est}}). \tag{6.6}$$

The task in this chapter is to find a combination of a measurement update rule u and estimator $\hat{\rho}^{\text{est}}$ with average expected loss as small as possible.

6.1.3 A-Optimality Criterion

The A-optimality criterion[1] is a measurement update criterion based on the asymptotic theory of statistical parameter estimation in Sect. 3.2.2 [1, 2].

Let $H(s)$ denote a positive semidefinite matrix on the parameter space, and we consider a quadratic loss function $\Delta(s, s') := (s' - s) \cdot H(s)(s' - s)$. In A-optimality criterion, we try to choose the next measurement as making the expected loss $\bar{\Delta}_N$ smaller. For the purpose, the Cramér-Rao bound is used. The Cramér-Rao inequality, Eq. (3.27), is valid not only for IID, but also for adaptive designs of experiments. From Eq. (3.27), we can derive the following inequality for any asymptotically unbiased estimator s^{est}:

$$\lim_{N\to\infty} N\bar{\Delta}_N(u, s^{\text{est}}|s) \geq \lim_{N\to\infty} N\text{tr}[H(s)F(u^N, s)^{-1}], \tag{6.7}$$

[1] The "A" stands for "average-variance".

where $F(u^N, s)$ is the Fisher matrix of the probability distribution $\{p(D^N | u, s)\}_{D^N \in \mathscr{D}^N}$. Roughly speaking, Eq. (6.7) means that for sufficiently large N, the expected loss $\bar{\Delta}_N(u, s^{\text{est}}|s)$ is lower bounded by $\text{tr}[H(s)F(u^N, s)^{-1}]$. This is a bound independent of estimator, and as explained in Theorem 3.1, a maximum-likelihood estimator attains the equality of Eq. (6.7). Therefore for a given adaptive design u, the bound $\text{tr}[H(s)F(u^N, s)^{-1}]$ can be interpreted as the best performance in all asymptotically unbiased estimator when N is sufficiently large.

In A-optimality criterion, we wish to minimize the value of $\text{tr}[H(s)F(u^N, s)^{-1}]$. Suppose that we perform n trials and obtained the data sequence D^n. We would like to choose the POVM minimizing $\text{tr}[H(s)F(u^N, s)^{-1}]$ in $\mathscr{M}_{n+1}$ as the next, $(n+1)$-th, measurement. When we consider minimizing this function, there are two problems. In order to avoid them, we introduce two approximations. The first problem is that the minimized function depends on the true parameter s. Of course the true parameter is unknown in parameter estimation problems, and we must use a dummy estimate in the update criterion, $s_n^{\text{dmm}}(D^n)$, instead. The dummy estimator s^{dmm} for the measurement update is not necessarily the same as the actual estimator s^{est}. The second problem is that unlike the independent and identically distributed (i.i.d.) measurement case, calculation of the Fisher matrix in the adaptive case requires summing over an exponential amount of data, and is computationally intensive. To avoid this problem, we approximate the sum over all possible measurements by that over only those measurements that have been performed:

$$F(u^{n+1}, s) \approx \tilde{F}(u^{n+1}, s|D^n) := \sum_{i=1}^{n+1} F(A_i, s|D^{i-1}), \tag{6.8}$$

where A_i is the parameter corresponds to the i-th POVM $\boldsymbol{\Pi}_i = u_i(D^{i-1})$ and

$$F(A_i, s|D^{i-1}) := \sum_{x_i \in \mathscr{X}_i} \frac{\nabla_s p(x_i|A_i, s) \nabla_s^T p(x_i|A_i, s)}{p(x_i|A_i, s)}. \tag{6.9}$$

The matrix $F(A_i, s)$ is the Fisher matrix for the i-th measurement probability distribution $\{p(x_i|A_i, s)\}_{x_i \in \mathscr{X}_i}$, and $\tilde{F}(u^{n+1}, s|D^n)$ is the sum of the Fisher matrices from the first to the $(n+1)$-th trial. $\tilde{F}(u^{n+1}, s|D^n)$ is called the conditional Fisher matrix. Instead of minimizing $\text{tr}[H(s)F(u^{n+1}, s)^{-1}]$, we consider the minimization of $\text{tr}[H(s_n^{\text{dmm}})\tilde{F}(u^{n+1}, s_n^{\text{dmm}})^{-1}]$. It is known that the convergence of $\text{tr}[H(s)\tilde{F}(u^{N+1}, s)^{-1}]$ to 0 at $N = \infty$ is part of a sufficient condition for the convergence of a maximum likelihood estimator [3], and this justifies the use of this second approximation. We explain the relationship between the conditional and unconditional Fisher matrices with respect to the estimator's convergence in Sect. A.1.3. Let $\mathscr{A}_n$ denote the set of parameters corresponding POVMs in $\mathscr{M}_n$ $(n = 1, \ldots, N)$. We use the same notation of update criterion for parameter A as that of the POVM. After making these two approximations, we define the A-optimality criterion in parameter representation as

$$
\begin{aligned}
A_{n+1}^{\text{A-opt}} &:= u_{n+1}^{\text{A-opt}}(D^n) \\
&= \underset{A_{n+1}\in\mathscr{A}_{n+1}}{\text{argmin}} \ \text{tr}[H(s_n^{\text{dmm}})\tilde{F}(u^{n+1}, s_n^{\text{dmm}}|D^n)^{-1}] \qquad (6.10) \\
&= \underset{A\in\mathscr{A}_{n+1}}{\text{argmin}} \, \text{tr}\Big[H(s_n^{\text{dmm}})\Big\{\sum_{i=1}^{n} F(A_i, s_n^{\text{dmm}}|D^{i-1}) + F(A, s_n^{\text{dmm}}|D^n)\Big\}^{-1}\Big] \qquad (6.11)
\end{aligned}
$$

Finding $A_{n+1}^{\text{A-opt}}$ is a nonlinear minimization problem with high computational cost in general. In the next section, we derive the analytic solution of Eq. (6.11) in the one-qubit case. This analytic solution reduces the computational cost significantly. We also perform numerical simulation and compare the performance of two A-optimality criteria with two nonadaptive designs of experiments.

6.2 Adaptive One-Qubit State Estimation

In Sect. 6.2, we consider adaptive quantum estimation using the A-optimality criteria in one-qubit system and analyze the estimation errors numerically. In Sect. 6.2.1, we explain the terminology and notation for adaptive estimation. In Sect. 6.2.2, we explain our theoretical and numerical results. We derive the explicit form of the analytic solution of the A-optimality criterion, and show numerical plots of pointwise and average expected losses. These plots indicate that the A-optimality criterion gives us more accurate estimates than standard state tomography and an other adaptive design of experiments proposed. In Sect. 6.2.3, we discuss an experimental implementation of A-optimality criterion in one-photon polarization system and the generalization to the higher ($d > 3$) dimensional systems.

6.2.1 Estimation Setting

We consider a one-qubit mixed state estimation problem, so that $\mathscr{O} = \mathscr{S}(\mathbb{C}^2)$. We assume that the true state is mixed, i.e., the Bloch vector $\boldsymbol{s}$ is strictly in the interior of the Bloch Ball, $B^o := \{\boldsymbol{s} \in \mathbb{R}^3 | \|\boldsymbol{s}\|_2 < 1\}$ with $\mathscr{O}$. From a theoretical viewpoint, this is to avoid the possible divergence of the Fisher matrix (in general, when the true density matrix is non-fullrank, the Fisher matrix can diverge). From an experimental viewpoint, this restriction can be justified because in real experiment there are effects of environmental systems and we cannot prepare strictly pure states. Suppose that we can choose any rank-1 projective measurement in each trial. Let $\boldsymbol{\Pi}(\boldsymbol{a}) = \{\hat{\Pi}_x(\boldsymbol{a})\}_{x=\pm 1}$ denote the POVM corresponding to the projective measurement onto the $\boldsymbol{a}$-axis ($\boldsymbol{a} \in \mathbb{R}^3$, $\|\boldsymbol{a}\|_2 = 1$), whose elements can be represented as

$$\hat{\Pi}_{\pm 1}(\boldsymbol{a}) = \frac{1}{2}(\hat{\ } \pm \boldsymbol{a}\cdot\boldsymbol{\sigma}). \qquad (6.12)$$

This is the Bloch parametrization of projective measurements. We identify the set of parameters

$$\mathscr{A} = \{\boldsymbol{a} \in \mathbb{R}^3 \mid \|\boldsymbol{a}\| = 1\} \tag{6.13}$$

with the measurement class

$$\mathscr{M} = \{\text{All rank-1 projective measurements on a one-qubit system}\}. \tag{6.14}$$

Unlike in quantum tomography, the performed measurements are not fixed and we cannot introduce relative frequencies in adaptive estimation. Therefore, linear estimators cannot be defined. Norm-minimization estimators are defined by using a linear estimator and they also cannot be defined. In numerical simulation, as the dummy and actual estimators both, we use a maximum likelihood estimators $\hat{\rho}^{\mathrm{ML}}$ and $\boldsymbol{s}^{\mathrm{ML}}$ defined as

$$\hat{\rho}_n^{\mathrm{ML}} := \underset{\hat{\rho}' \in \mathscr{O}}{\operatorname{argmax}}\, p(D^n | u, \hat{\rho}'), \tag{6.15}$$

$$\boldsymbol{s}_n^{\mathrm{ML}} := \underset{\boldsymbol{s}' \in B}{\operatorname{argmax}}\, p(D^n | u, \hat{\rho}(\boldsymbol{s}')), \tag{6.16}$$

for $n = 1, \ldots, N$.

For our loss functions, we use both the squared Hilbert-Schmidt distance Δ^{HS2}

$$\Delta^{\mathrm{HS2}}(\boldsymbol{s}, \boldsymbol{s}') := \frac{1}{2} \mathrm{Tr}[(\rho(\boldsymbol{s}) - \rho(\boldsymbol{s}'))^2] \tag{6.17}$$

$$= \frac{1}{4}(\|\boldsymbol{s} - \boldsymbol{s}'\|_2)^2, \tag{6.18}$$

and the infidelity Δ^{IF} in Eqs. (5.20) and (5.22). We note that the Hilbert-Schmidt distance coincides with the trace distance in a one-qubit system. The asymptotic behavior of the average expected fidelity $\bar{\Delta}_N^{\mathrm{IFave}}$ is known in the one-qubit state estimation case [4–6]. The probability measure used for calculating this average is the Bures distribution, $d\mu(\boldsymbol{s}) = \frac{1}{\pi^2}(1 - \|\boldsymbol{s}\|^2)^{-1/2} d\boldsymbol{s}$. If we limit our available measurements to be sequential and independent (i.e., nonadaptive), $\bar{\Delta}_N^{\mathrm{IFave}}$ behaves at best as $O(N^{-3/4})$ [4, 5]. On the other hand, if we are allowed to use adaptive, separable, or collective measurements, $\bar{\Delta}_N^{\mathrm{IFave}}$ can behave as $O(N^{-1})$ [6]. In [4–6], the coefficient of the dominant term in the asymptotic limit is also derived.

In Sect. 6.2.2.2, we show numerical results. It is shown that the average expected infidelity of an A-optimal scheme behaves as $O(N^{-1})$, illustrating that the A-optimality criterion is indeed making use of adaptation to outperform nonadaptive schemes.

6.2.2 Results and Analysis

As explained in Sect. 6.2.1, we consider the A-optimality criterion for one-qubit state estimation using projective measurements. In Sect. 6.2.2.1 we give the analytic solution, and in Sect. 6.2.2.2 we show the results of our numerical simulations.

6.2.2.1 Analytic Solution for A-Optimality in One-Qubit State Estimation

First, we give the explicit form of the Fisher matrix for projective measurements. The probability distribution for the rank-1 projective measurement $\boldsymbol{\Pi}(\boldsymbol{a})$ is given by

$$p(\pm 1|\boldsymbol{a}, \boldsymbol{s}) = \frac{1}{2}(1 \pm \boldsymbol{s} \cdot \boldsymbol{a}), \tag{6.19}$$

and the Fisher matrix is

$$\begin{aligned} F(\boldsymbol{a}, \boldsymbol{s}) &= \frac{\nabla_s p(+1|\boldsymbol{a}, \boldsymbol{s}) \nabla_s^T p(+1|\boldsymbol{a}, \boldsymbol{s})}{p(+1|\boldsymbol{a}, \boldsymbol{s})} + \frac{\nabla_s p(-1|\boldsymbol{a}, \boldsymbol{s}) \nabla_s^T p(-1|\boldsymbol{a}, \boldsymbol{s})}{p(-1|\boldsymbol{a}, \boldsymbol{s})} &(6.20) \\ &= \frac{\boldsymbol{a}\boldsymbol{a}^T}{1 - (\boldsymbol{a} \cdot \boldsymbol{s})^2}. &(6.21) \end{aligned}$$

The conditional Fisher matrix up to the n-th trial is given by

$$\tilde{F}(u^n, \boldsymbol{s}|D^n) = \sum_{i=1}^{n} \frac{\boldsymbol{a}_i \boldsymbol{a}_i^T}{1 - (\boldsymbol{a}_i \cdot \boldsymbol{s})^2}, \tag{6.22}$$

where $\boldsymbol{a}_i = u_i(D^{i-1})$. In this case, Eq. (6.11) is rewritten in the Bloch vector representation as

$$\begin{aligned} \boldsymbol{a}_{n+1}^{\text{A-opt}} &:= \text{argmin}_{\boldsymbol{a} \in \mathscr{A}} \operatorname{tr}\Big[H(\boldsymbol{s}_n^{\text{dmm}}) \Big\{ \tilde{F}(u^n, \boldsymbol{s}_n^{\text{dmm}}|D^n) + F(\boldsymbol{a}, \boldsymbol{s}_n^{\text{dmm}}) \Big\}^{-1} \Big] &(6.23) \\ &= \text{argmin}_{\boldsymbol{a} \in \mathscr{A}} \operatorname{tr}\Big[H(\boldsymbol{s}_n^{\text{dmm}}) \Big\{ \tilde{F}(u^n, \boldsymbol{s}_n^{\text{dmm}}|D^n) + \frac{\boldsymbol{a}\boldsymbol{a}^T}{1 - (\boldsymbol{a} \cdot \boldsymbol{s}_n^{\text{ddm}})^2} \Big\}^{-1} \Big] &(6.24) \end{aligned}$$

We present the analytic solution of Eq. (6.23) in the form of the following theorem.

Theorem 6.1. *Given a sequence of data $D^n = \{(\boldsymbol{a}_1, x_1), \ldots, (\boldsymbol{a}_n, x_n)\}$, the n-th dummy estimate $\boldsymbol{s}_n^{\text{dmm}}$, and a real positive matrix H, the A-optimal POVM Bloch vector is given by*

$$\boldsymbol{a}_{n+1}^{\text{A-opt}} = \frac{B_n \boldsymbol{e}_{\min}(C_n)}{\| B_n \boldsymbol{e}_{\min}(C_n) \|_2}, \tag{6.25}$$

where

$$B_n = \sqrt{\tilde{F}(u^n, s_n^{\mathrm{dmm}}|D^n) H(s_n^{\mathrm{dmm}})^{-1} \tilde{F}(u^n, s_n^{\mathrm{dmm}}|D^n)}, \tag{6.26}$$

$$C_n = B_n (I - s_n^{\mathrm{dmm}} s_n^{\mathrm{dmm}T} + \tilde{F}(u^n, s_n^{\mathrm{dmm}}|D^n)^{-1}) B_n, \tag{6.27}$$

$\boldsymbol{e}_{\min}(C_n)$ *is the eigenvector of the matrix* C_n *corresponding to the minimal eigenvalue, and* I *is the identity matrix on the parameter space.*

We give the proof of Theorem 6.1 in Sect. A.1.2.

In Eq. (6.27), the inverse of the matrix $\tilde{F}(u^n, s_n^{\mathrm{dmm}}|D^n)$ appears. In the proof of Theorem 6.1, the invertibility of $\tilde{F}(u^n, s_n^{\mathrm{dmm}}|D^n)$ is assumed. The invertibility of $\tilde{F}(u^n, s_n^{\mathrm{dmm}}|D^n)$ is equivalent to the condition that $\boldsymbol{a}^n = \{\boldsymbol{a}_1, \ldots, \boldsymbol{a}_n\}$ is a basis of $\mathbb{R}^3$. When we choose the second and third measurements, $\tilde{F}(u^1, s_1^{\mathrm{dmm}}|D^1)$ and $\tilde{F}(u^2, s_2^{\mathrm{dmm}}|D^2)$ are not invertible. Thus the update scheme does not apply to these steps, and the choices are arbitrary. One simple choice is to perform $\hat{\sigma}_1$-, $\hat{\sigma}_2$-, and $\hat{\sigma}_3$-projective measurements at the first, second and third trials respectively, and this can be shown to satisfy Theorem 6.1 as follows. The choice of the first measurement is always arbitrary, and we choose $\boldsymbol{a}_1 = (1, 0, 0)^T$, a $\hat{\sigma}_1$-projective measurement. Then for any true Bloch vector $\boldsymbol{s}$ the rank of $\tilde{F}(u^1, s_1^{\mathrm{dmm}}|D^1)$ is 1, and if we interpret the inverse matrix in Eq. (6.27) as a generalized inverse matrix, C_1 is a rank 1 matrix with minimal eigenvalues 0. The supports of $\tilde{F}(u^1, s_1^{\mathrm{dmm}}|D^1)$, B_1, and C_1 are the span of $\{\boldsymbol{a}_1\}$. Therefore $B_1 \boldsymbol{e}_{\min}(C_1)$ is an arbitrary vector in the 2-dimensional space spanned by $(0, 1, 0)^T$ and $(0, 0, 1)^T$, and we choose $\boldsymbol{a}_2 = (0, 1, 0)^T$. Then using the same logic, the third measurement is fixed to $\boldsymbol{a}_3 = (0, 0, 1)^T$.

From the explicit formulae of the squared Hilbert-Schmidt distance and infidelity in Eqs. (6.18) and (5.22), we have

$$\Delta^{\mathrm{HS2}}(\boldsymbol{s}, \boldsymbol{s}') = (\boldsymbol{s}' - \boldsymbol{s})^T \frac{1}{4} I (\boldsymbol{s}' - \boldsymbol{s}), \tag{6.28}$$

$$\Delta^{\mathrm{IF}}(\boldsymbol{s}, \boldsymbol{s}') = (\boldsymbol{s}' - \boldsymbol{s})^T \frac{1}{4}\left(I + \frac{\boldsymbol{s}\boldsymbol{s}^T}{1 - \|\boldsymbol{s}\|_2^2}\right)(\boldsymbol{s}' - \boldsymbol{s}) + O(\|\boldsymbol{s}' - \boldsymbol{s}\|_2^3). \tag{6.29}$$

Therefore when we use the Hilbert-Schmidt distance as our loss function, we substitute $H^{\mathrm{HS2}}(\boldsymbol{s}) := \frac{1}{4} I$ and $H^{\mathrm{HS}}(\boldsymbol{s})^{-1} = 4I$ into Eqs. (6.26), and (6.27). Then we have

$$B_n = \tilde{F}(u^n, s_n^{\mathrm{dmm}}|D^n), \tag{6.30}$$

$$C_n = \tilde{F}(u^n, s_n^{\mathrm{dmm}}|D^n)(I - s_n^{\mathrm{dmm}} s_n^{\mathrm{dmm}T}) \tilde{F}(u^n, s_n^{\mathrm{dmm}}|D^n) + \tilde{F}(u^n, s_n^{\mathrm{dmm}}|D^n), \tag{6.31}$$

and we do not need to explicitly calculate the inverse or square root matrices for A-optimality. On the other hand, when our loss function is the infidelity, we must use $H^{\mathrm{IF}}(\boldsymbol{s}) := \frac{1}{4}\left(I + \frac{\boldsymbol{s}\boldsymbol{s}^T}{1-\|\boldsymbol{s}\|_2^2}\right)$ and $H^{\mathrm{IF}}(\boldsymbol{s})^{-1} = 4(I - \boldsymbol{s}\boldsymbol{s}^T)$. This is based on the quadratic approximation of infidelity.

6.2.2.2 Numerical Simulation

We performed Monte Carlo simulations of the following four experimental designs described in detail below; A-optimal adaptive scheme for the squared Hilbert-Schmidt distance, the same for infidelity, XYZ repetition, and uniformly random selection.

A-optimality for the squared Hilbert-Schmidt distance is the adaptive scheme defined by Eq. (6.23) with $H = H^{\mathrm{HS}}$. Similarly, A-optimality for the infidelity is that with $H = H^{\mathrm{IF}}$. As explained in the previous subsection, the choice of measurement Bloch vectors at the first and second trials is arbitrary; we choose $\boldsymbol{a}_1 = (1, 0, 0)^T$ and $\boldsymbol{a}_2 = (0, 1, 0)^T$, i.e., at the first trial we perform the projective measurement of $\hat{\sigma}_1$, and that of $\hat{\sigma}_2$ at the second—the third trial is automatically the projective measurement of $\hat{\sigma}_3$, corresponding to $\boldsymbol{a}_3 = (0, 0, 1)^T$. The XYZ repetition scheme is nonadaptive, in which we repeat the measurements of $\hat{\sigma}_1$, $\hat{\sigma}_2$, and $\hat{\sigma}_3$, corresponding to standard quantum state tomography. Uniformly random selection is also nonadaptive, where at each trial we choose the next measurement direction randomly on the Bloch surface, according to the SO(3) Haar measure. For consistency with the other three schemes, we fix the first, second and third measurements to be the projective measurements of $\hat{\sigma}_1, \hat{\sigma}_2, \hat{\sigma}_3$, respectively, and randomly select directions from the fourth trial on.

We choose a maximum likelihood estimator in all four experimental designs. It is known that the estimators minimizing $\bar{\Delta}^{\mathrm{HS2\text{-}ave}}$ and $\bar{\Delta}^{\mathrm{IF\text{-}ave}}$ are Bayesian estimators [4, 7], but the integrations necessary for Bayesian estimation take too much computation time. For the two A-optimality criteria, we choose both the real and the dummy estimators to be maximum likelihood, $\boldsymbol{s}^{\mathrm{est}} = \boldsymbol{s}^{\mathrm{dmm}} = \boldsymbol{s}^{\mathrm{ML}}$. We used a Newton-Raphson method to solve the (log-)likelihood equation and the completely mixed state $\boldsymbol{s} = \boldsymbol{0}$ as the initial point of the iterative method. When a search point came out of the Bloch sphere during the procedure, we chose the previous point (included in the sphere) as the estimate.

In the following subsections, we show the plots for two loss functions; the squared Hilbert-Schmidt distance Δ^{HS2} and infidelity Δ^{IF}. The average expected losses $\bar{\Delta}_N^{\mathrm{ave}}$ and pointwise expected losses $\bar{\Delta}_N$ are shown. In the both plots, the line styles are fixed as follows:

- Solid (black) line for A-optimality for the squared Hilbert-Schmidt distance (AHS)
- Dashed (red) line for A-optimality for the infidelity (AIF)
- Chain (blue) line for XYZ repetition (XYZ)
- Dotted (green) line for Uniformly random selection (URS).

1. **Average expected losses**

We analyse the average behaviour of the estimation errors over the Bloch sphere. The integration for averaging is approximated by a Monte Carlo routine, and the

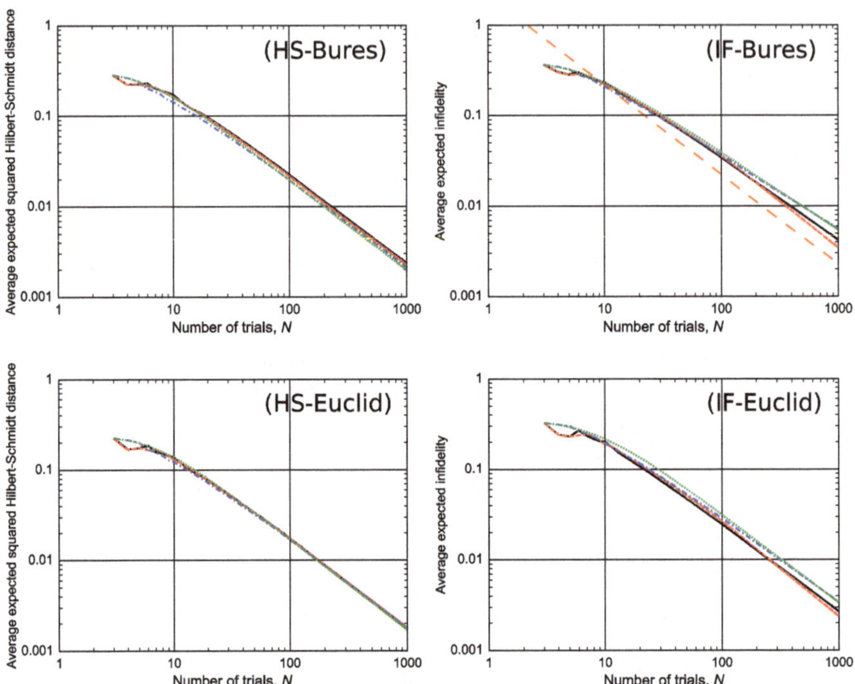

Fig. 6.2 Average expected loss $\bar{\Delta}_N^{\text{ave}}(u^N, \boldsymbol{s}^{\text{ML}})$ against the number of measurement trials N: *(HS-Bures)* $\bar{\Delta}_N^{\text{HS2-ave}}$ integrated via the Bures distribution μ_{Bures}, *(HS-Euclid)* $\bar{\Delta}_N^{\text{HS2-ave}}$ integrated via the Euclidean distribution $\mu_{\text{Euclid}}(\boldsymbol{s}) = 3/4\pi$, *(IF-Bures)* $\bar{\Delta}_N^{\text{IF-ave}}$ integrated via μ_{Bures}, and *(IF-Euclid)* $\bar{\Delta}_N^{\text{IF-ave}}$ integrated via μ_{Euclid}. The line styles are fixed as follows: *solid* (*black*) *line* for A-optimality for the squared Hilbert-Schmidt distance (AHS), *dashed* (*red*) *line* for A-optimality for the infidelity (AIF), *chain* (*blue*) *line* for XYZ repetition (XYZ), and *dotted* (*green*) *line* for Uniformly random selection (URS). The *dashed spaced* (*orange*) *line* in (*IF-Bures*) is the bound of separable (including adaptive) schemes derived in [6]. The number of measurement trials $N_{\max}$ is 1,000, the number of sequences used for the calculation of the statistical expectation values N_{mean} is 1,000, and the number of sample points used for the Monte Carlo integration N_{MC} is 3,200. (Reproduced from Ref. [24] with permission)

statistical expectation is approximated by an arithmetic mean using pseudo-random numbers.

Figure 6.2 shows the average expected loss functions $\bar{\Delta}_N^{\text{ave}}(u, \boldsymbol{s}^{\text{ML}})$ against the number of trials N (the horizontal and vertical axes are both logarithmic scale):

- (HS-Bures) $\bar{\Delta}_N^{\text{HS2-ave}}$ integrated via the Bures distribution μ_{Bures}
- (HS-Euclid) $\bar{\Delta}_N^{\text{HS2-ave}}$ integrated via the Euclidean distribution $\mu_{\text{Euclid}}(\boldsymbol{s}) = 3/4\pi$
- (IF-Bures) $\bar{\Delta}_N^{\text{IF-ave}}$ integrated via μ_{Bures}
- (IF-Euclid) $\bar{\Delta}_N^{\text{IF-ave}}$ integrated via μ_{Euclid}.

Figure 6.2 (HS-Bures) and (HS-Euclid) shows that the estimation errors of the four experimental designs are almost equivalent from the viewpoint of the squared

Hilbert-Schmidt distance. As depicted in (HS-Bures), the estimation errors of the two A-optimality schemes are slightly larger than the other nonadaptive schemes; as we show in the pointwise analysis below, this gap decreases as N becomes larger. On the other hand, Fig. 6.2 (IF-Bures) and (IF-Euclid) show the explicit gap between the adaptive and nonadaptive schemes. The gradients of the curves begin to differentiate from around $N = 100$, and as depicted in (IF-Bures), the gradients of XYZ and URS are almost $-3/4$ around $N = 1{,}000$. This means that the average expected infidelity behaves as $O(N^{-3/4})$ and is consistent with the result of the asymptotic analysis presented in [4]. On the other hand, the gradients of AHS and AIF are greater than the nonadaptive limit $-3/4$, indicating that AHS and AIF make good use of adaptive resources. Around $N = 1{,}000$ the gradient of AIF is almost -1, which is the bound for adaptive experimental designs [6].

Let us compare the estimation errors of A-optimality and the HH12 criteria explained in Sect. A.1.1 7. From Fig. 6.2 (IF-Bures), the average expected infidelity of AHS and AIF are 4.2×10^{-3} and 3.5×10^{-3} at $N = 1{,}000$. On the other hand, the corresponding amount for the HH12 criterion can be estimated roughly from Fig. 2a in [8] to be 7.0×10^{-3}. This implies that for one-qubit state estimation, the average expected infidelity of the A-optimality criterion is about two-times smaller than that of Eq. (6.44), at least around $N = 1{,}000$.

2. **Pointwise expected losses**

Next, we analyse the behaviour of the estimation errors at several true Bloch vectors $\boldsymbol{s}$. Figure 6.3 shows the pointwise expected loss functions $\bar{\Delta}_N(u, \boldsymbol{s}^{\mathrm{ML}}|\boldsymbol{s})$ against the number of trials N (the horizontal and vertical axes are both logarithmic scale): (HS-P1), (HS-P2), and (HS-P3) are plots of the expected squared Hilbert-Schmidt distances for $\boldsymbol{s}$ given by $(r, \theta, \phi) = (0, 0, 0),\ (0.99, 0, 0),\ (0.99, \pi/4, \pi/4)$, and (IF-P1), (IF-P2), and (IF-P3) are the expected infidelities for the same three true states, respectively.

As depicted in (HS-P1) and (IF-P1), the estimation errors of all four schemes are almost equivalent for the completely mixed state, $\boldsymbol{s} = \boldsymbol{0}$. As the Bloch radius r becomes larger, the differences between the four schemes become clearer. Figure 6.3 (HS-P2) and (HS-P3) are the plots of the expected squared Hilbert-Schmidt distances at a high purity point, $r = 0.99$. In the region of $N = 10$ to around 7,000, the squared Hilbert-Schmidt error of the two adaptive schemes is larger than that of the two nonadaptive schemes. In particular, the error of AHS is larger that that of AIF; this might seem strange, but in the region of $N \geq 7{,}000$, the error of AHS becomes smaller than that of AIF, indeed it eventually becomes the smallest of the four schemes. We believe that there are two reasons for A-optimality's large error for small N. First, the A-optimality criterion is based on an asymptotic theory of statistical estimation. When the number of measurement trials N is small, the Cramér-Rao bound is not necessary suitable for characterizing estimation errors. Second, it uses a dummy estimator in the measurement update. When $n = 1, \ldots, N$ is small, $\boldsymbol{s}_n^{\mathrm{dmm}} (= \boldsymbol{s}_n^{\mathrm{ML}}$ in our numerical simulation) is not a good estimate, and thus the choice of the next measurements can be unreliable. Of course, when N becomes sufficiently large, both of these problems are alleviated.

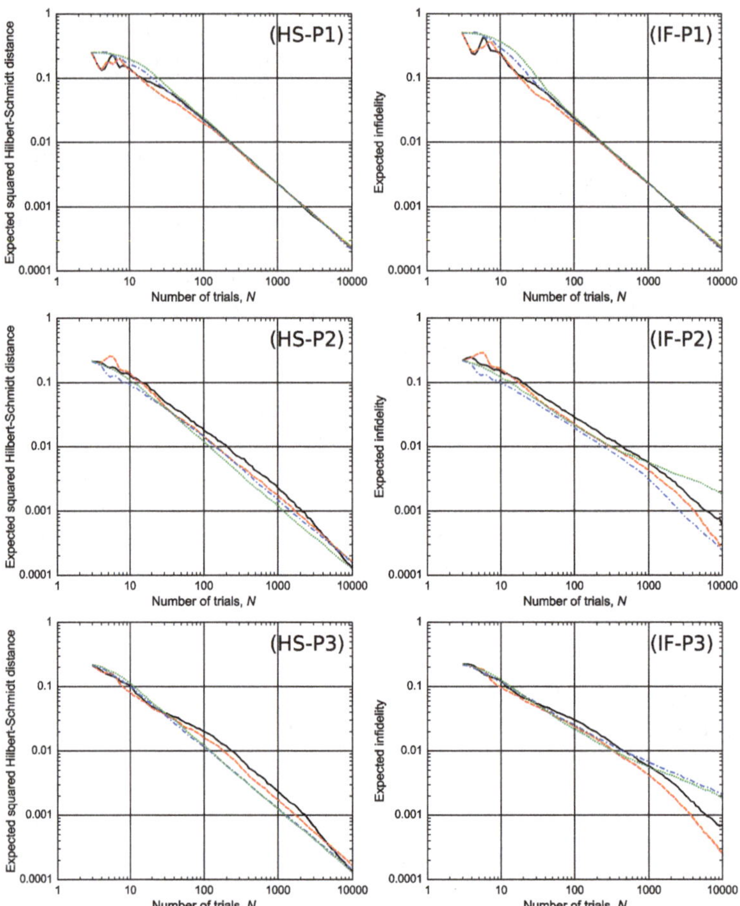

Fig. 6.3 Pointwise expected loss $\bar{\Delta}_N(u, \mathbf{s}^{\mathrm{ML}}|\mathbf{s})$ against the number of trials N: (*HS-P1*), (*HS-P2*), and (*HS-P3*) are the expected squared Hilbert-Schmidt distances for $\mathbf{s}$ given by $(r, \theta, \phi) = (0, 0, 0)$, $(0.99, 0, 0)$, $(0.99, \pi/4, \pi/4)$, and (*IF-P1*), (*IF-P2*), and (*IF-P3*) are the expected infidelities for the same three true states, respectively.The line styles are fixed as follows: *solid* (*black*) *line* for A-optimality for the squared Hilbert-Schmidt distance (AHS), *dashed* (*red*) *line* for A-optimality for the infidelity (AIF), *chain* (*blue*) *line* for XYZ repetition (XYZ), and *dotted* (*green*) *line* for Uniformly random selection (URS). The number of measurement trials N_{max} is 10,000, and the number of sequences used for the calculation of statistical expectation values N_{mean} is 1,000. (Reproduced from Ref. [24] with permission)

The gap between the estimation errors of adaptive and nonadaptive schemes becomes smaller as N becomes larger in (HS-P2) and (HS-P3), while it grows in (IF-P2) and (IF-P3). Only the XYZ scheme changes dramatically between (IF-P2) and (IF-P3); the other three schemes do not because AHS, AIF, and URS are invariant under rotation of the true Bloch vector (for very small N, there are differences, and these are because the first three measurements are fixed to $\hat{\sigma}_1, \hat{\sigma}_2, \hat{\sigma}_3$-projective measurements and not rotationally invariant). Figure 6.3 (IF-P2) is the case in which the directions of the measurement and the true Bloch vector are matched (to (0, 0, 1)). In this case, XYZ is the best scheme, exhibiting the smallest estimation error. Around $N = 10{,}000$, the estimation error of AIF becomes as small as that of XYZ. That of AHS is smaller than URS, but larger than the other two schemes. We believe that this is because the selected Hessian matrix H^{HS2} used in the update routine is unsuitable for the loss function Δ^{IF} in (IF-P2) (and (IF-P3)). Figure 6.3 (IF-P3) is the case in which the directions of the measurement and the true Bloch vector are the most discrepant (for a fixed purity). In this case, the estimation errors of XYZ and URS are almost the same and behave as $O(N^{-1/2})$, and those of the adaptive schemes are smaller than those of the nonadaptive ones. When we consider the whole Bloch sphere, of course the cases in which the direction of XYZ measurements and the Bloch vector are matched are few, and therefore the average expected infidelities of AHS and AIF are smaller than those of XYZ and URS. This also indicates that the adaptive schemes have better worst-case performance (lower $\bar{\Delta}_N^{\max}$) than the nonadaptive schemes.

3. **Purity dependence**

Figure 6.4 shows the purity dependence of the average expected infidelity at $N = 1{,}000$. The average is taken over all directions θ and ϕ for each Bloch radius r. It indicates that the average expected infidelities of the two adaptive schemes are smaller than those of the two nonadaptive schemes. The appearance of peaks for XYZ and URS is discussed in Sect. A.1.4.

4. **Measurement sequences**

Figure 6.5 is a plot of the measurement Bloch vectors at $N = 100$ (left column), 1,000 (middle column), and 10,000 (right column) for 900 runs. The true state is $(r, \theta, \phi) = (0.99, \pi/4, \pi/4)$, and the upper three subplots are AHS while the lower three are AIF. Figure 6.5 shows that the measurement Bloch vectors are clustered around the true state, with some interesting behaviour at $N = 10{,}000$. In (AHS-10,000), the measurement directions are clustered very narrowly at the true state and also around the great circle that it defines. In (AIF-10,000), on the other hand, the directions are clustered widely around the true state. This is due to the difference between the loss functions employed in the update routine, namely squared Hilbert-Schmidt distance in the former and infidelity in the latter. We mention that for a completely mixed true state, the measurement Bloch vectors are distributed randomly on the Bloch sphere for large N.

6.2.3 Discussion

In this subsection, we discuss an experimental implementation of the A-optimal design of experiments and the generalization of A-optimality criterion to higher dimensional systems.

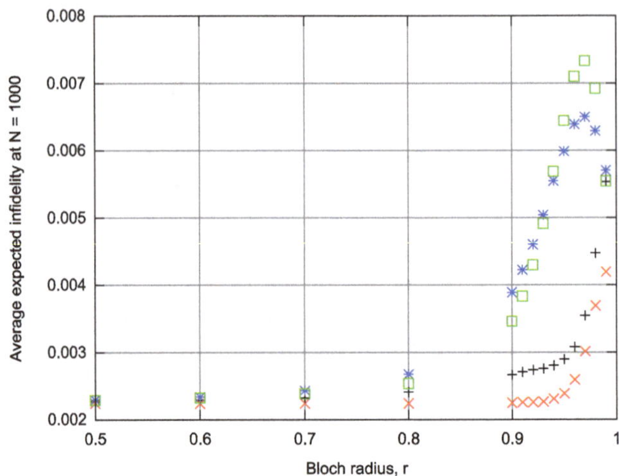

Fig. 6.4 Purity dependence of average expected infidelity at $N = 1{,}000$: *Cross* (*black*) for AHS, *saltire* (*red*) for AIF, *asterisk* (*blue*) for XYZ, and *square* (*green*) for URS. The number of sequences used for the calculation of the statistical expectation values N_{mean} is 1,000, and the number of sample points used for the Monte Carlo integration N_{MC} is 500 for each Bloch radius r. (Reproduced from Ref. [24] with permission)

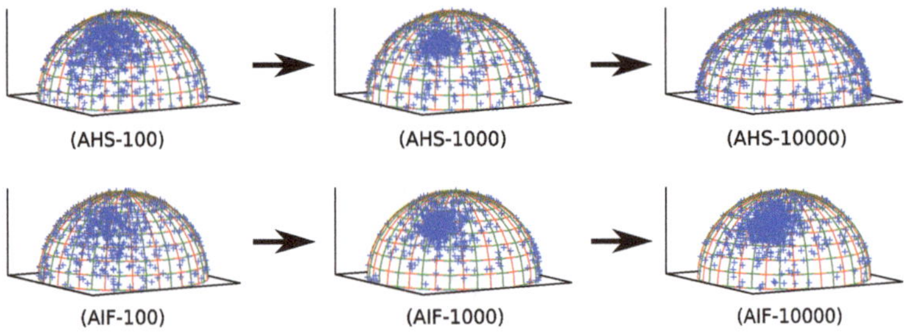

Fig. 6.5 Distribution of measurement Bloch vectors at $N = 100$ (*left column*), 1,000 (*middle column*), and 10,000 (*right column*) for 900 runs: The true state is $(r, \theta, \phi) = (0.99, \pi/4, \pi/4)$. The *upper* three plots (*AHS*-100), (*AHS*-1,000), and (*AHS*-10,000) are AHS while the *lower* three (*AIF*-100), (*AIF*-1,000), and (*AIF*-10,000) are AIF. (Reproduced from Ref. [24] with permission)

6.2.3.1 Implementation

There are two main issues when considering the practical implementation of an adaptive scheme, namely the ease with which measurement updates can be made in the apparatus, and the time required to compute those updates. In quantum optics, projective measurements and single qubit rotations are standard tools in quantum information processing experiments. Figure 6.6 illustrates a simple implementation example for a one photon polarization system. In this regard, the first issue is not a problem—in general, of course it will depend on the experimental state of the art.

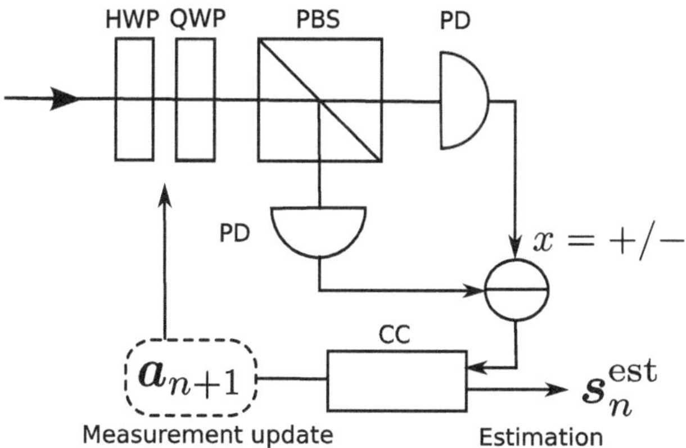

Fig. 6.6 An implementation of adaptive projective measurements for single photon polarization qubits in quantum optics: *HWP* and *QWP* are half and quarter wave plates, *PBS* is a polarization beam splitter, *PD* are photodetectors, and *CC* denotes classical computation. The direction of the projective measurement are adapted by changing the waveplate angles. (Reproduced from Ref. [24] with permission)

6.2.3.2 Generalization to Higher Dimensional Systems

In order to compare the performance of the A-optimality criterion to the other update schemes, we have considered one-qubit states as the estimation objective. Current and future quantum information processing is concerned with higher dimensional estimation objectives, not only states but also processes. In one-qubit state estimation, we can reduce the computational cost for A-optimality by using the analytic solution of Theorem 6.1, but as we see in Sect. A.1.2, the techniques used to derive that solution depend on the properties of one-qubit states and projective measurements. A-optimality in higher dimensional systems will need a new solution, or must deal with the increasing complexity of the nonlinear minimization problem. One possible approach is to place constraints on the measurement class $\mathscr{M}_n$. Instead of considering a continuous set of measurement candidates, we could consider a discrete set. One expects that the resulting discrete minimization problem would be much simpler. If the number of discrete measurement candidates is too small however, the estimation error could be worse than standard quantum tomography. The relation between the reduction in computational cost and the (probable) increase in estimation error by introducing such discrete minimization is an open problem.

6.3 History

As explained in Sect. 6.1, adaptive designs of experiments are characterized by a measurement update criterion. Previously proposed update criteria include those based on asymptotic statistical estimation theory (Fisher information) [9–11], direct calculations of the estimates expected to be obtained in the next measurement [12, 13], mutually unbiased basis [14], as well as Bayesian estimators and Shannon entropy [8, 12, 15]. The details of these update criteria are explained in Sect. A.1.1. Theoretical investigations report that some of the proposed update criteria give more precise estimates than nonadaptive experimental designs, and an experimental implementation of the update criterion proposed in [12] and in [11] has been performed in an ion trap system [16] and single photon system [17], respectively.

6.4 Summary

In this chapter, we considered adaptive experimental design and applied a measurement update method known in statistics as the A-optimality criterion to one-qubit mixed state estimation using arbitrary rank-1 projective measurements. In Sect. 6.1, we explained basic concepts and terminologies in adaptive design of experiments and gave the definition and idea of A-optimality criterion. In Sect. 6.2, we applied the A-optimality criterion and analyzed the performance. We derived an analytic solution of the A-optimality update procedure in this case. The analytic solution reduces the complexity of measurement updates considerably. Our analytic solution is applicable to any case in which the loss function can be approximated by a quadratic function to least order. We performed Monte Carlo simulation of this and several nonadaptive schemes in order to compare the behaviour of estimation errors for a finite number of measurement trials. We compared the average and pointwise expected squared Hilbert-Schmidt distance and infidelity of the following four measurement update criteria: A-optimality for the squared Hilbert-Schmidt distance (AHS), A-optimality for the infidelity (AIF), repetition of three orthogonal projective measurements (XYZ), and uniformly random selection of projective measurements (URS). The numerical results showed that AHS and AIF give more precise estimates than URS and XYZ which corresponds to standard quantum tomography with respect to expected infidelity. In Sect. 6.3, we reviewed the history of adaptive design of experiments in quantum estimation.

Appendix

A.1 Supplement for Chap. 6

In this section, we give supplements for this chapter. In Sect. A.1.1, we briefly review some adaptive measurement update criteria proposed in the literature. In Sect. A.1.2 we give the proof of Theorem 6.1. In Sect. A.1.3, we explain the relation between conditional and unconditional Fisher matrices. We also show that unconditional Fisher matrix, rather than conditional one, plays an essential role in adaptive design of experiments with a simple example. In Sect. A.1.4, we explain the origin of a peak appearing in Fig. 6.4.

A.1.1 Survey of Proposed Update Criteria

We briefly review some adaptive measurement update criteria proposed in the literature, using our terminology and notation introduced in Sect. 6.1.

1. **Two-step adaptation criterion**

Before explaining update criteria that are performed at each and every trial, such as A-optimality, we briefly review a simpler update criterion. The two-step adaptation criterion requires the measurement update only once during a measurement sequence. We have

$$u_{n+1}(D^n) = \begin{cases} \boldsymbol{\Pi}_{1\text{st}} & \text{if } n < N_{1\text{st}} \\ \boldsymbol{\Pi}_{2\text{nd}} & \text{if } n \geq N_{1\text{st}} \end{cases}. \tag{6.32}$$

Thus, for all trials up to and including trial $N_{1\text{st}}$ a fixed POVM $\boldsymbol{\Pi}_{1\text{st}}$ is performed, and an estimate is calculated from the obtained data. Using that data we choose a new POVM $\boldsymbol{\Pi}_{2\text{nd}}$ for the remaining $N_{2\text{nd}}(= N - N_{1\text{st}})$ copies. In [6, 18–20], two-step adaptation criteria are used to prove mathematically an asymptotic bound for weighted mean squared errors in one-qubit state estimation. In [21, 22], some numerical results are shown for a few two-step adaptation schemes.

2. **N88 criterion**

In [9–11], an update criterion based on the Cramér-Rao inequality is proposed. The update criterion is given by

$$u_{n+1}(D^n) = \underset{A \in \mathscr{A}_{n+1}}{\operatorname{argmin}} \operatorname{tr}[H(\boldsymbol{s}_n^{\text{dmm}}) F(A, \boldsymbol{s}_n^{\text{dmm}})^{-1}]. \tag{6.33}$$

The difference from the A-optimality criterion is that in Eq. (6.33) the Fisher information matrix used in the update does not take into account all $n+1$ measurements, but about only the $(n+1)$-th measurement. The advantage of course is that this reduces the computational cost of updates. The disadvantage is that when $\mathscr{M}_n$ $(n = 1, 2, \ldots)$ consists of informationally incomplete POVMs, as is the case in most experiments, the estimates cannot converge to the true state. As explained in Sect. 6.2.1, in Sect. 6.2 $\mathscr{M}_n$ is restricted to rank-1 projective measurements, and in this setting Eq. (6.33) does not work well.

3. **FKF00 criteria**

In [12], two update criteria are proposed.

(i) The first criterion is based on the Shannon entropy of the estimated measurement probability distribution, and is given by

$$u_{n+1}(D^n) = \underset{\boldsymbol{\Pi} \in \mathscr{M}_{n+1}}{\operatorname{argmax}} \left\{ - \sum_{x \in \mathscr{X}_{n+1}} p(x|\boldsymbol{\Pi}, \hat{\rho}_n^{\text{dmm}}(D^n)) \ln p(x|\boldsymbol{\Pi}, \hat{\rho}_n^{\text{dmm}}(D^n)) \right\}. \tag{6.34}$$

(ii) The second criterion uses an additional dummy estimator $\hat{\rho}^{\mathrm{dmm2}}$ such that

$$\left(u_{n+1}(D^n), \hat{\rho}_n^{\mathrm{dmm2}}(D^n)\right) = \underset{(\boldsymbol{\Pi}, \hat{\rho}') \in \mathscr{M}_{n+1} \times \mathscr{O}}{\mathrm{argmax}} \left\{ \sum_{x \in \mathscr{X}_{n+1}} p(x|\boldsymbol{\Pi}, \hat{\rho}_n^{\mathrm{dmm}}(D^n)) \Delta(\hat{\rho}_{n+1}^{\mathrm{dmm}}(D^{n+1}), \hat{\rho}') \right\}. \tag{6.35}$$

Numerical simulation is performed for the case where $\mathscr{O}$ is the set of one-qubit pure states and $\mathscr{M}_n$ is the set of projective measurements, while $\hat{\rho}^{\mathrm{est}}$ is a biased maximum likelihood estimator, $\hat{\rho}^{\mathrm{est}}$ is a Bayesian estimator up to $N = 60$. Average (not expected) infidelity is used as the evaluation function.

4. **HF08 criterion**

In [13], an update criterion given by

$$u_{n+1}(D^n) = \underset{\boldsymbol{\Pi} \in \mathscr{M}_{n+1}}{\mathrm{argmax}} \left\{ \int_{\mathscr{O}} d\mu(\hat{\rho}) \sum_{x \in \mathscr{X}_{n+1}} p(x|\boldsymbol{\Pi}, \hat{\rho}) \Delta(\hat{\rho}_{n+1}^{\mathrm{dmm}}(D^{n+1}), \rho) \right\}, \tag{6.36}$$

is proposed. A numerical simulation is performed in [13], where the setting is that $\mathscr{O}$ is the set of one-qubit pure states, $\mathscr{M}$ is a set of parity measurements using an ancilla system, and $\hat{\rho}^{\mathrm{est}}$ and $\hat{\rho}^{\mathrm{dmm}}$ are maximum likelihood estimators. The behavior of the average expected fidelity is numerically analyzed up to $N = 20$.

5. **HF11 criterion**

An update criterion proposed in [14] is given by

$$u_{n+1}(D^n) = \underset{\boldsymbol{\Pi} \in \mathscr{M}_{n+1}}{\mathrm{argmax}} \left(- \sum_{x \in \mathscr{X}_{n+1}} \sum_{i=1}^{n} \mathrm{Tr}[\hat{\Pi}_{i,x_i} \hat{\Pi}_x] \ln \mathrm{Tr}[\hat{\Pi}_{i,x_i} \hat{\Pi}_x] \right), \tag{6.37}$$

and the estimator is defined as

$$\hat{\rho}_n^{\mathrm{est}}(D^n) = \underset{\hat{\rho}' \in \mathscr{O}}{\mathrm{argmax}}\, \mathrm{Tr}[\hat{\rho}' \hat{\rho}_0(D^n)], \tag{6.38}$$

$$\hat{\rho}_0(D^n) = \frac{1}{n} \sum_{i=1}^{n} \hat{\Pi}_{i,x_i}. \tag{6.39}$$

In the numerical simulations, the estimation setting is such that $\mathscr{O}$ is the set of pure states on d-dimensional Hilbert space $\mathscr{H}$, and $\mathscr{M}_n$ is the set of projective measurements on $\mathscr{H}$. Numerical simulations of average expected fidelity are shown for $d = 2, 4, 6, 8$, and 13, all up to $N = 50$.

6. **FF00 criterion**

In [15], an update criterion based on Bayesian estimation and Shannon entropy is proposed. Let $P(\hat{\rho})$ denote a prior distribution on $\mathscr{O}$. The update criterion is

$$u_{n+1}(D^n) = \underset{\boldsymbol{\Pi}\in\mathscr{M}_{n+1}}{\mathrm{argmax}} \left\{ \sum_{x\in\mathscr{X}_{n+1}} p^{\mathrm{ave}}(x|\boldsymbol{\Pi}, D^n) \int_{\hat{\rho}\in\mathscr{O}} d\mu(\hat{\rho}) P(\hat{\rho}|D^{n+1}) \ln \frac{P(\hat{\rho}|D^{n+1})}{P(\hat{\rho}|D^n)} \right\} \tag{6.40}$$

$$= \underset{\boldsymbol{\Pi}\in\mathscr{M}_{n+1}}{\mathrm{argmax}} \left\{ - \int_{\hat{\rho}\in\mathscr{O}} d\mu(\hat{\rho}) P(\hat{\rho}|D^n) \ln P(\hat{\rho}|D^n) \right. \\ \left. + \sum_{x\in\mathscr{X}_{n+1}} p^{\mathrm{ave}}(x|\boldsymbol{\Pi}, D^n) \int_{\mathscr{O}} d\mu(\hat{\rho}) P(\hat{\rho}|D^{n+1}) \ln P(\hat{\rho}|D^{n+1}) \right\} \tag{6.41}$$

where

$$p^{\mathrm{ave}}(x|\boldsymbol{\Pi}, D^n) := \int_{\mathscr{O}} d\mu(\hat{\rho}) P(\hat{\rho}|D^n) p(x|\boldsymbol{\Pi}, \hat{\rho}), \tag{6.42}$$

$$P(\hat{\rho}|D^n) := \frac{P(\hat{\rho}) p(D^n|u, \hat{\rho})}{\int_{\mathscr{O}} d\mu(\hat{\rho}') P(\hat{\rho}') p(D^n|u, \hat{\rho}')}. \tag{6.43}$$

In [15], the case in which $\mathscr{O}$ is the set of one-qubit mixed states and $\mathscr{M}$ is the set of projective measurements is numerically analyzed up to $N = 50$. The evaluation function used is the average (not expected) infidelity.

7. **HH12 criterion**

In [8], an update criterion given by

$$u_{n+1}(D^n) = \underset{\boldsymbol{\Pi}\in\mathscr{M}_{n+1}}{\mathrm{argmax}} \left\{ - \sum_{x\in\mathscr{X}_{n+1}} p^{\mathrm{ave}}(x|\boldsymbol{\Pi}, D^n) \ln p^{\mathrm{ave}}(x|\boldsymbol{\Pi}, D^n) \right. \\ \left. + \int_{\mathscr{O}} d\mu(\hat{\rho}) P(\hat{\rho}|D^n) \sum_{x\in\mathscr{X}_{n+1}} p(x|\boldsymbol{\Pi}, \hat{\rho}) \ln p(x|\boldsymbol{\Pi}, \hat{\rho}) \right\}, \tag{6.44}$$

is proposed, where Eqs. (6.42) and (6.43) have been used. From a simple calculation, we can see that the criteria defined in Eq. (6.41) and in Eq. (6.44) are equivalent. This criterion involves an integration which requires high computational cost. In [8], a special technique for calculating the integral, called a sequential importance sampling method, is used in order to reduce that computational cost. The authors performed numerical simulation for the case in which $\mathscr{O}$ is the set of one-qubit mixed states and $\mathscr{M}_n$ are projective measurements up to $N = 10^4$. They also considered the case in which $\mathscr{O}$ is the set of two-qubit states and $\mathscr{M}$ are a set of mutually unbiased bases, a set of pairwise Pauli measurements, and a set of separable measurements up to

$N = 10^5$. The evaluation function is the average expected infidelity, and it is shown that their scheme is more precise than standard quantum tomography. In Sect. 6.2.2.2, we point out that our numerical results for one-qubit show that A-optimality gives even more precise estimates than those given by Eq. (6.44), at least from $N = 100$ to 1,000.

A.1.2 Proof of Theorem 6.1

We give the proof of Theorem 6.1. First, we introduce a lemma about matrix inverses [23].

Lemma 6.1. *Let V denote a $k \times k$ invertible matrix. Let us consider a matrix $W = V + \boldsymbol{v}\boldsymbol{v}^T$, where $\boldsymbol{v}$ is a k-dimensional vector. If W is not singular, then*

$$W^{-1} = V^{-1} - \frac{V^{-1}\boldsymbol{v}\boldsymbol{v}^T V^{-1}}{1 + \boldsymbol{v}^T V^{-1}\boldsymbol{v}}. \tag{6.45}$$

By substituting $\boldsymbol{v} = \boldsymbol{a}/\sqrt{1 - (\boldsymbol{a} \cdot \boldsymbol{s})}$ into Eq. (6.45) (in our case $k = 3$ and $V = \tilde{F}$), we obtain

$$\{V + F(\boldsymbol{a}, \boldsymbol{s})\}^{-1} = V^{-1} - \frac{V^{-1}\boldsymbol{a}\boldsymbol{a}^T V^{-1}}{1 - (\boldsymbol{a} \cdot \boldsymbol{s})^2 + \boldsymbol{a}^T V^{-1}\boldsymbol{a}}, \tag{6.46}$$

and

$$\mathrm{tr}[H(\boldsymbol{s})\{V + F(\boldsymbol{a}, \boldsymbol{s})\}^{-1}] = \mathrm{Tr}[H(\boldsymbol{s})V^{-1}] - \frac{\boldsymbol{a}^T V^{-1} H(\boldsymbol{s}) V^{-1}\boldsymbol{a}}{1 - (\boldsymbol{a} \cdot \boldsymbol{s})^2 + \boldsymbol{a}^T V^{-1}\boldsymbol{a}}. \tag{6.47}$$

The first term of the RHS in Eq. (6.47) is independent of $\boldsymbol{a}$ and therefore we obtain

$$\begin{aligned} &\operatorname*{argmin}_{\boldsymbol{a} \in \mathscr{A}} \mathrm{tr}[H(\boldsymbol{s})\{V + F(\boldsymbol{a}, \boldsymbol{s})\}^{-1}] \\ &= \operatorname*{argmax}_{\boldsymbol{a} \in \mathscr{A}} \frac{\boldsymbol{a}^T V^{-1} H(\boldsymbol{s}) V^{-1}\boldsymbol{a}}{\boldsymbol{a}^T (I - \boldsymbol{s}\boldsymbol{s}^T + V^{-1})\boldsymbol{a}} \qquad (6.48) \\ &= \operatorname*{argmin}_{\boldsymbol{a} \in \mathscr{A}} \frac{\boldsymbol{a}^T (I - \boldsymbol{s}\boldsymbol{s}^T + V^{-1})\boldsymbol{a}}{\boldsymbol{a}^T V^{-1} H(\boldsymbol{s}) V^{-1}\boldsymbol{a}}, \qquad (6.49) \end{aligned}$$

where we used the relation $1 = \boldsymbol{a}^T I \boldsymbol{a}$. Let us introduce a vector

$$\boldsymbol{b} := \frac{\sqrt{V^{-1}H(\boldsymbol{s})V^{-1}}\boldsymbol{a}}{\|\sqrt{V^{-1}H(\boldsymbol{s})V^{-1}}\boldsymbol{a}\|_2}. \tag{6.50}$$

Note that $\boldsymbol{b}$ and $\boldsymbol{a}$ take values in the same set, so that the vector $\boldsymbol{a}$ can be represented in terms of $\boldsymbol{b}$ as

$$\boldsymbol{a} = \frac{\sqrt{VH(s)^{-1}V}\boldsymbol{b}}{\|\sqrt{VH(s)^{-1}V}\boldsymbol{b}\|_2}. \tag{6.51}$$

Then the minimization function is represented by using $\boldsymbol{b}$ as

$$\frac{\boldsymbol{a}^T(I - \boldsymbol{s}\boldsymbol{s}^T + V^{-1})\boldsymbol{a}}{\boldsymbol{a}^T V^{-1} H(s) V^{-1}\boldsymbol{a}} = \boldsymbol{b}^T \sqrt{VH(s)^{-1}V}(I - \boldsymbol{s}\boldsymbol{s}^T + V^{-1})\sqrt{VH(s)^{-1}V}\boldsymbol{b}. \tag{6.52}$$

The vector $\boldsymbol{b}$ minimizing Eq. (6.52) is the eigenvector with the minimal eigenvalue of the matirx

$$C := \sqrt{VH(s)^{-1}V}(I - \boldsymbol{s}\boldsymbol{s}^T + V^{-1})\sqrt{VH(s)^{-1}V}, \tag{6.53}$$

i.e., $\boldsymbol{b} = \boldsymbol{e}_{\min}(C)$. By substituting $V = \tilde{F}$ and $\boldsymbol{s} = \boldsymbol{s}_n^{\mathrm{dmm}}$ into Eqs. (6.51) and (6.53), we obtain Theorem 6.1. □

A.1.3 Conditional Fisher matrices

In this section we explain the relation between conditional and unconditional Fisher matrices. From a simple calculation, we can obtain

$$F(u^N, \boldsymbol{s}) = \sum_{D^{N-1} \in \mathscr{D}^{n-1}} p(D^{N-1}|u, \boldsymbol{s})\tilde{F}(u^N, \boldsymbol{s}|D^{N-1}). \tag{6.54}$$

This is the reason why $\tilde{F}$ is called the conditional Fisher matrix of F. In statistical parameter estimation theory, it is known that the divergence of the conditional Fisher matrix ($\tilde{F}(u^N, \boldsymbol{s}|D^{N-1}) \to \infty$ as $N \to \infty$) almost everywhere in $\mathscr{D}^{N-1}$ is part of a sufficient condition for the convergence (known as *strong consistency* in statistics) of a maximum-likelihood estimator [3]. If we assume that the other elements of the set of sufficient conditions are satisfied, the divergence of the conditional Fisher matrix is sufficient for the convergence of a maximum-likelihood estimator. In this case, from Eq. (6.54), the unconditional Fisher matrix also diverges ($F(u^N, \boldsymbol{s}) \to \infty$), and this is equivalent to the condition that $\mathrm{tr}[F(u^N, \boldsymbol{s})^{-1}] \to 0$. Therefore, the divergence of the unconditional Fisher matrix is a necessary condition for the convergence of a maximum-likelihood estimator. The divergence of $F(u^N, \boldsymbol{s})$ is, however, not sufficient for the convergence of a maximum-likelihood estimator.

We illustrate this with a simple example. Suppose that our estimation objective is $\mathscr{O} = \mathscr{S}(\mathbb{C}^2)$. At the first trial, we perform a POVM $\boldsymbol{\Pi} = \{\hat{\Pi}_{\mathrm{T}}, \hat{\Pi}_{\mathrm{F}}\}$, where $\hat{\Pi}_{\mathrm{T}} = \hat{\Pi}_{\mathrm{F}} = \frac{1}{2}\hat{}$. We obtain the outcome T and F both with $1/2$ probability. When we obtain

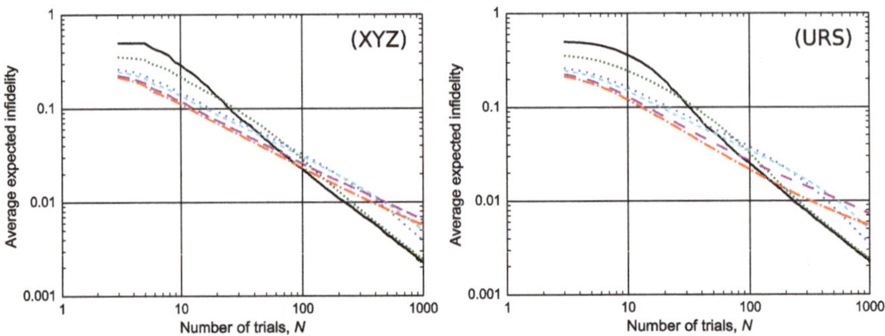

Fig. 6.7 Purity dependence of average expected infidelity of XYZ and URS schemes: The average is taken over all directions θ and ϕ for each Bloch radius r. Average expected infidelity of XYZ repetition (*left*) and URS (*right*) for different Bloch radii. *Solid line* (*black*): $r = 0$, *dotted line* (*green*): $r = 0.7$, *dotted spaced line* (*blue*): $r = 0.9$, *dashed line* (*light blue*): $r = 0.93$, *dashed spaced line* (*purple*): $r = 0.97$, and *dotted dashed line* (*red*): $r = 0.99$. The number of sequences used for the calculation of the statistical expectation values N_{mean} is 1,000, and the number of sample points used for the Monte Carlo integration N_{MC} is 500 for each Bloch radius r. (Reproduced from Ref. [24] with permission)

an outcome T at the first measurement, we perform standard quantum tomography for the rest of all the trials. In this case, the maximum-likelihood estimate converges to the true state, and the conditional Fisher matrix $\tilde{F}(u^N, \boldsymbol{s}|D^{N-1})$ whose D^{N-1} includes $x_1 = \text{T}$ diverges. Let $\tilde{F}(u^N, \boldsymbol{s}|\text{T})$ denote the conditional Fisher matrix. On the other hand, when we obtain F in the first measurement, we repeat the same POVM $\boldsymbol{\Pi}$ for the remaining trials. Let $\tilde{F}(u^N, \boldsymbol{s}|\text{F})$ denote the conditional Fisher matrix whose D^{N-1} includes $x_1 = \text{F}$. In this case, no estimator converges to the true state because the POVM $\boldsymbol{\Pi}$ does not give us any information (the probability distribution is (1/2, 1/2), independent of the true state). Then we obtain $\tilde{F}(u^N, \boldsymbol{s}|\text{F}) = 0$. The unconditional Fisher matrix is calculated as

$$
\begin{aligned}
F(u^N, \boldsymbol{s}) &= \frac{1}{2}\tilde{F}(u^N, \boldsymbol{s}|\text{T}) + \frac{1}{2}\tilde{F}(u^N, \boldsymbol{s}|\text{F}) && (6.55)\\
&= \frac{1}{2}\tilde{F}(u^N, \boldsymbol{s}|\text{T}) && (6.56)\\
&\to \infty, && (6.57)
\end{aligned}
$$

i.e., the unconditional Fisher matrix $F(u^N, \boldsymbol{s})$ diverges even though no estimator converges to the true state with probability 1/2. Therefore the divergence of $F(u^N, \boldsymbol{s})$ is necessary, but not sufficient for the convergence of a maximum-likelihood estimator.

As we can see from the above example, in adaptive experimental designs, the essential characteristic of the scheme is not the unconditional Fisher matrix but the conditional Fisher matrices. In order to make a maximum-likelihood estimator converge, we need to design an experiment such that almost all (not necessarily strictly all) the conditional Fisher matrices diverge. From this point of view, the approximation Eq. (6.8) lies at the heart of adaptive experimental designs.

A.1.4 Purity Dependence of XYZ and URS Schemes

In Fig. 6.4 it is shown that the average expected infidelities of XYZ and URS at $N = 1{,}000$ have a peak around $r = 0.97$. Here we explain the origin of the peak. Figure 6.7 is a plot of average expected infidelity for six Bloch radii (purities) r. We choose six purities from the fourteen purities in Fig. 6.4 to make things easier to see. The average is taken over all directions θ and ϕ for each Bloch radius r; (XYZ) is for XYZ and (URS) is for URS. Roughly speaking, the plots can be interpreted as straight lines with different slopes and y-intercepts on a log-log scale. As the purity (r) increases, two things occur: (i) the slope of the curves becomes less steep, and (ii) the y-intercept decreases. This change of the slope is caused by the bias of a maximum-likelihood estimator, which has been explained in Sect. 5.2.3. The turning point of the slope is characterized by N^* defined in Eq. (5.88). At $N = 1{,}000$, these two effects combine in such a way as to create a peak in the estimation error around $r = 0.97$.

References

1. S. Watanabe, K. Hagiwara, S. Akaho, Y. Motomura, K. Fukumizu, M. Okada, M. Aoyagi, *Theory and Implimentation of Learning Systems* (Morikita, Japan, 2005)
2. F. Pukelsheim, *Optimal Design of Experiments*. Classics in Applied Mathematics (SIAM, Philadelphia, 2006)
3. P. Hall, C.C. Heyde, *Martingale Limit Theory and Its Application*. Probability and mathematical statistics (Academic Press, New York, 1980)
4. E. Bagan, M. Baig, R. Muñoz-Tapia, A. Rodriguez, Phys. Rev. A **69**, 010304(R) (2004). doi:10.1103/PhysRevA.69.010304
5. E. Bagan, M.A. Ballester, R.D. Gill, A. Monras, R. Muñoz-Tapia, Phys. Rev. A **73**, 032301 (2006). doi:10.1103/PhysRevA.73.032301
6. E. Bagan, M.A. Ballester, R.D. Gill, R. Muñoz-Tapia, O. Romero-Isart, Phys. Rev. Lett. **97**, 130501 (2006). doi:10.1103/PhysRevLett.97.130501
7. R. Blume-Kohout, New J. Phys. **12**, 043034 (2010). doi:10.1088/1367-2630/12/4/043034
8. F. Huszár, N.M.T. Houlsby, Phys. Rev. A **85**, 052120 (2012). doi:10.1103/PhysRevA.85.052120
9. H. Nagaoka, in *Proceedings of 12th Symposium on Information Theory and Its Applications*, (1989), p. 577
10. H. Nagaoka, in *Asymptotic Theory of Quantum Statistical Inference: Selected Papers*, Chap. 10, World Scientific, 2005, ed. by M. Hayashi
11. A. Fujiwara, J. Phys. A: Math. Gen. **39**, 12489 (2006). doi:10.1088/0305-4470/39/40/014
12. D.G. Fischer, S.H. Kienle, M. Freyberger, Phys. Rev. A **61**, 032306 (2000). doi:10.1103/PhysRevA.61.032306
13. C.J. Happ, M. Freyberger, Phys. Rev. A **78**, 064303 (2008). doi:10.1103/PhysRevA.78.064303
14. C.J. Happ, M. Freyberger, Eur. Phys. J. D **64**, 579 (2011). doi:10.1140/epjd/e2011-20367-9
15. D.G. Fischer, M. Freyberger, Phys. Lett. A **273**, 293 (2000). doi:10.1016/S0375-9601(00)00513-2
16. T. Hannemann, D. Reiss, C. Balzer, W. Neuhauser, P.E. Toschek, C. Wunderlich, Phys. Rev. A **65**, 050303 (2002). doi:10.1103/PhysRevA.65.050303
17. R. Okamoto, M. Iefuji, S. Oyama, K. Yamagata, H. Imai, A. Fujiwara, S. Takeuchi, Phys. Rev. Lett. **109**, 130404 (2012). doi:10.1103/PhysRevLett.109.130404

18. M. Hayashi, K. Matsumoto, in *RIMS Kôkyûroku*, vol. 1055 (Research Institute for Mathematical Sciences, Kyoto University, Kyoto, 1998), p. 96 (The original version is in Japanese. An English version is available in the arXiv.)
19. M. Hayashi, K. Matsumoto, in *Asymptotic Theory of Quantum Statistical Inference: Selected Papers*, Chap. 13, World Scientific, 2005, ed. by M. Hayashi
20. R.D. Gill, S. Massar, Phys. Rev. A **61**, 042312 (2000). doi:10.1103/PhysRevA.61.042312
21. J. Řeháček, B.G. Englert, D. Kaszlikowski, Phys. Rev. A **70**, 052321 (2004). doi:10.1103/PhysRevA.70.052321
22. D. Petz, K.M. Hangos, L. Ruppert, in *Quantum bio-informatics, QP-PQ: Quantum Probability and White Noise Analysis*, vol. 21, ed. by L. Accardi, L. Accardi, M. Ohya (2007), p. 247
23. R.A. Horn, C.R. Johnson, *Matrix Analysis* (Cambridge University Press, New York, 1985)
24. T. Sugiyama, P.S. Turner, M. Murao, Phys. Rev. A **85**, 052107 (2012). doi:10.1103/PhysRevA.85.052107

Chapter 7
Summary and Outlook

In this thesis we analyzed statistical estimation errors in the test of Bell-type correlations and quantum tomography for finite samples.

- Chapter 2 Preliminary 1: Quantum Mechanics and Quantum Estimation—Background and Problems in Quantum Estimation

  In Chap. 2, we explained the postulates of quantum mechanics used in quantum information and reviewed quantum estimation. We particularly emphasized the importance of finite sample analysis of estimation errors in the test of Bell-type correlations and quantum tomography.
- Chapter 3 Preliminary 2: Mathematical statistics—Basic Concepts and Theoretical Tools for Finite Sample Analysis

  In Chap. 3, we explained the fundamental concepts and known results in statistical estimation. These are necessary for analyzing estimation errors in quantum estimation. In Sect. 3.1, we explained the basic concepts and terminology in probability theory and the theory of statistical parameter estimation. Two figures of merit for estimation errors analyzed in this thesis, expected loss and error probability, were explained. In Sect. 3.2, we explained known results for the behavior of the expected loss and error probability for infinitely large samples. For both figures of merit, it was explained that the optimal rate or coefficient of decrease is attainable by a maximum-likelihood estimator, and when the experimental design is identical and independently prepared, a linear estimator can also attain the optimality. In Sect. 3.3, we explained known results for the behavior of the arithmetic mean of a random variable for finite samples. These are elementary and well known, but useful for analyzing estimation errors of quantum estimation with finite samples.
- Chapter 4 Evaluation of Estimation Precision in Test of Bell-type Correlations

  In Chap. 4, we analyzed the behavior of expected losses and error probabilities in a test of Bell-type correlations. In Sect. 4.1, we explained a fundamental concept in quantum non-locality, namely quantum entanglement. In Sect. 4.2, we explained the CHSH inequality and its relation to entanglement. In Sect. 4.3, we analyze the estimation errors in the test of the CHSH inequality. We chose a linear estimator, and derived some explicit forms of upper bounds on the expected loss and error

T. Sugiyama, *Finite Sample Analysis in Quantum Estimation*, Springer Theses,
DOI: 10.1007/978-4-431-54777-8_7,

probability. These explicit forms are directly applicable to evaluating the validity of the violation of the inequality in a CHSH experiment.

- Chapter 5 Evaluation of Estimation Precision in Quantum Tomography

  In Chap. 5, we analyzed the behavior of expected losses and error probabilities in quantum state tomography with finite data. In Sect. 5.1, we explained the estimation setting in quantum state tomography. We proposed a new estimator called an extended norm-minimization estimator, and focused these of three estimators; extended linear, extended norm-minimization, and maximum-likelihood estimators. In Sects. 5.2 and 5.3, we derived some functions upper-bounding the expected losses and error probabilities for those three estimators. The derived functions are valid for any true estimation objects in arbitrary finite dimensional systems, and these are applicable to tomographic experiments.

  As explained in Sect. 5.1.2.2, in many quantum state tomography experiments, the final goal is to verify a successful preparation of a specific state. The quantity that we should evaluate is the difference between what we prepared and what we want to prepare. In many experiments, however, this quantity has not been able to be evaluated because we do not know what we prepared, and instead the evaluated quantity is the difference between what we estimate and what we want. By using the functions upper-bounding the expected losses and error probabilities for the extended ℓ_2-norm-minimization estimator derived in Sects. 5.2 and 5.3, we can evaluate the difference between what we prepared and what we want to prepare. By using these functions, we can evaluate the performance of an experimental state preparation for a quantum information protocol. This is the largest contribution to quantum information science in this thesis.

  The functions upper-bounding expected losses in Sect. 5.2 involve a maximization, and so the explicit forms were not derived. In order to evaluate the values, we need to solve the maximization problems. To derive the explicit functional form is an open problem. The author feels there is a possibility of improvement for the functions upper-bounding error probabilities in Sect. 5.3. In the derivation of the functions, we used four inequalities, and the functions are probably not tight upper-bounds, so finding better bounds is also an open problem. Finally, the functions upper-bounding the expected losses and error probabilities hold only for finite-dimensional Hilbert space. There are important estimation problems for infinite-dimensional Hilbert space, for example, homodyne tomography in quantum optics [1]. To derive upper-bounds for the infinite case is also an important open problem.

- Chapter 6 Improvement of Estimation Precision by Adaptive Design of Experiments

  In Chap. 6, we considered an adaptive design of experiment in one-qubit state estimation. We focused a measurement update criterion, the A-optimality criterion known in the classical theory of experimental design. In Sect. 6.1, we explained basic concepts and terminologies in adaptive design of experiments and gave the definition and idea of A-optimality criterion. In Sect. 6.2, we applied the A-optimality criterion for a one-qubit state estimation and analyzed the performance. We derived the analytic solution of the update criterion in the case for using rank-1 projective measurements. By using the analytic solution, we performed a Monte Carlo simulation and numerically evaluated the average and pointwise expected losses. Our numerical results indicated that the A-optimal design of experiments gives us more precise estimates than the standard quantum state tomography and other adaptive design of experiments proposed in the literature.

As shown numerically in Sect. 6.2, by using the A-optimality criterion, we can improve estimation precision compared to standard state tomography and an adaptive design of experiments in one-qubit state estimation. Our analytical results are, however, valid only for one-qubit systems. The author expects that estimation errors of the A-optimality criterion are smaller than state tomography also in higher dimensional systems. In general, adaptive design of experiments requires the solution of an optimization problem for updating measurements, and this has computational cost. In one-qubit systems, we can reduce the computational cost by using our analytic solution for the A-optimality criterion, but in higher dimensional systems it is not guaranteed that analytic solutions can be derived. It is a very difficult problem to derive analytic solutions in higher dimensional systems, and we consider it better to numerically solve the optimization problem. In (classical) machine learning theory, numerical methods for efficiently solving optimization problems are proposed [2]. We expect that computational cost can be reduced in higher dimensional systems by combining such methods with quantum estimation theory.

In this thesis, we established a theory of expected loss and error probability in quantum estimation for finite data sets. We believe that our results contribute to the developments of quantum information science, and thus of physics. We hope that experimentalists might use our results to analyze their experimental data.

References

1. A.L. Lvovsky, M.G. Raymer, Rev. Mod. Phys. **81**, 299 (2009). doi:10.1103/RevModPhys.81.299
2. S. Watanabe, K. Hagiwara, S. Akaho, Y. Motomura, K. Fukumizu, M. Okada, M. Aoyagi, *Theory and Implimentation of Learning Systems* (Morikita Publishing Co., 2005). (in Japanese)

Curriculum Vitae

Dr. Takanori Sugiyama

Personal details

Affiliation	Institute for Theoretical Physics, Department of Physics, ETH Zurich, Wolfgang-Pauli-Strasse 27, CH-8093 Zurich, Switzerland
E-mail	sugiyama@itp.phys.ethz.ch

Education

04/2010–03/2013	Doctor of Science Department of Physics, The University of Tokyo, Japan Thesis title: *Finite sample analysis in quantum estimation*
04/2008–03/2010	Master of Science Department of Physics, The University of Tokyo, Japan Thesis title: *Analysis of quantum tomographic reconstruction schemes*
04/2004–03/2008	Bachelor of Science Department of Physics, Keio University, Japan Thesis title: *Quantum measurement theory and uncertainty relations*

T. Sugiyama, *Finite Sample Analysis in Quantum Estimation*, Springer Theses,
DOI: 10.1007/978-4-431-54777-8,

Academic positions

04/2013–present	Postdoctoral Researcher (JSPS Postdoctoral Fellow for Research Abroad) Institute for Theoretical Physics, Department of Physics ETH Zurich, Switzerland
04/2010–03/2013	JSPS Research Fellow for Young Scientists Department of Physics, Graduate School of Science The University of Tokyo, Japan

Awards

03/2013	Student Research Prize Graduate School of Science, The University of Tokyo, Japan

Zeitfracht Medien GmbH
Ferdinand-Jühlke-Straße 7
99095 Erfurt, Deutschland
produktsicherheit@kolibri360.de